LEADERSHIP AND FORCE DEVELOPMENT

UNITED STATES

AIR FORCE

Air Force Doctrine Document 1-1
18 February 2006

Interim Change to 18 February 2004 Version

This document complements related discussion found in Joint Publication 0-2, *Unified Action Armed Forces (UNAAF)*.

BY ORDER OF THE
SECRETARY OF THE AIR FORCE

AIR FORCE DOCTRINE DOCUMENT 1-1

18 FEBRUARY 2006

SUMMARY OF REVISIONS

This document was revised on 6 January 2006 to incorporate discussions on operational risk management as a responsibility of leadership (page 3) and language addressing longer term force sustainment efforts within force development (page 24). Paragraphs revised are annotated with a solid line in the right border.

Supersedes: AFDD 1-1, 18 Feb 04
OPR: HQ AFDC/DR (Mr Bob Christensen)
Certified by: AFDC/CC (Maj Gen Bentley B. Rayburn)
Pages: 88
Distribution: F
Approved by: JOHN P. JUMPER, General, USAF
                         Chief of Staff

FOREWORD

This document discusses leadership and force development principles and tenets that are experience-based and rooted in all levels of the Air Force. It emphasizes the leader's personal engagement; these are the leaders who make things better. Airmen should use it as guidance for fulfilling their assigned duties and leadership responsibilities.

The Air Force's core values are the foundation of leadership. *Integrity* is the basis of trust, and trust is the unbreakable bond that unifies leaders with their followers and commanders with their units. Trust makes leaders effective, and integrity underpins trust. *Service Before Self* is the essence of our commitment to the nation. Leaders who serve selflessly inspire support from everyone in their command and promote a spirit that binds organizations into an effective warfighting team. *Excellence in All We Do* is our commitment to the highest standards of service to our country. Leaders set the standard for excellence in their organizations.

Leaders do not appear fully developed out of whole cloth. A maturation must occur to allow the young leaders to grow into the responsibilities required of senior institutional leaders and commanders. This force development process provides leadership focus at all levels in an Airman's career. The expeditionary Air Force requires leaders who can take warfighting to the highest possible level of success in support of our national security objectives. Those leaders can only be created through an iterative process of development involving education, training, and expeditionary operations seasoned with experience and ongoing mentoring by more experienced leaders. The end result is an individual capable of successfully operating as a leader at all levels anywhere, anytime.

Force development takes individual capabilities and, through education, training, and experience, produces skilled, knowledgeable, and competent Airmen who can apply the best tools, techniques, and procedures to produce a required operational capability. We prepare Airmen for leadership by optimizing experiences and skills and by developing capabilities to meet any challenges.

This document is *the* Air Force statement of leadership principles and force development, enabled by education and training, providing a framework for action ensuring our Airmen can become effective leaders. Your personal leadership is the key to our Service's success in fulfilling its role in our system of national security.

JOHN P. JUMPER
General, USAF
Chief of Staff

TABLE OF CONTENTS

INTRODUCTION

PURPOSE

This Air Force Doctrine Document (AFDD) establishes doctrinal guidance for leadership and force development in the United States Air Force.

APPLICATION

This AFDD applies to all active duty, Air Force Reserve, Air National Guard, and civilian Air Force personnel.

The doctrine in this document is authoritative. Therefore, commanders need to consider the contents of this AFDD and the particular situation when accomplishing their missions. Airmen should read it, discuss it, and practice it.

SCOPE

This document provides guidance for Air Force leaders in fulfilling assigned missions. It ensures leaders at every echelon throughout the Air Force have a baseline for preparing themselves and their forces to conduct operations. This is essential for the success of the highly flexible and rapidly responsive operations in which the Air Force routinely engages either independently or as a component of a joint/multinational task force. Doctrine describes the proper use of air and space forces in military operations and serves as a guide for the exercise of professional judgment rather than a set of inflexible rules. It describes the Air Force's understanding of the best way to do the job to accomplish national objectives.

FOUNDATIONAL DOCTRINE STATEMENTS

Foundational doctrine statements are the basic principles and beliefs upon which AFDDs are built. Other information in the AFDDs expands on or supports these statements.

✪ Leadership is the art and science of influencing and directing people to accomplish the assigned mission.

✪ Leadership does not equal command, but all commanders should be leaders.

✪ The abilities of a leader, which are derived from innate capabilities and built from experience, education, and training, can be improved upon through deliberate development.

✪ The core values are a statement of those institutional values and principles of conduct that provide the moral framework within which military activities take place.

✪ The professional Air Force ethic consists of three fundamental and enduring values of integrity, service, and excellence.

✪ As leaders move into the most complex and highest levels of the Air Force, or become involved in the strategic arena, the ability to conceptualize and integrate becomes increasingly important.

✪ Leadership skills needed at successively higher echelons in the Air Force build on those learned at previous levels.

✪ Developing Airmen best happens through a deliberate process, one that aims to produce the right capabilities to meet the Air Force's operational needs.

✪ Force development is a series of experiences and challenges, combined with education and training opportunities that are directed at producing Airmen who possess the requisite skills, knowledge, experience, and motivation to lead and execute the full spectrum of Air Force missions.

✪ Development processes and systems take individual capabilities and, through education, training, and experience, produce skilled, knowledgeable, and competent Airmen who can apply the best tools, techniques, and procedures to produce a required operational capability.

✪ The Air Force prepares Airmen for leadership by optimizing experiences and skills to provide an effective understanding of the appropriate levels of the organization and by developing capabilities to meet those challenges.

✪ Education and training are critical components of the force development construct. Education and training represent a large investment of resources and are the primary tools in developing Airmen.

CHAPTER ONE

AIR FORCE LEADERSHIP

Our warriors are no longer limited to the people who fly the airplanes.... Our entire force is a warrior force. Being a warrior is not an AFSC (Air Force specialty code),...it's a condition of the heart.

—General John Jumper, Chief of Staff, United States Air Force (CSAF), 2003

LEADERSHIP DEFINED

Leadership is the art and science of influencing and directing people to accomplish the assigned mission. This highlights two fundamental elements of leadership: (1) the mission, objective, or task to be accomplished, and (2) the people who accomplish it. All facets of Air Force leadership should support these two basic elements. Effective leadership transforms human potential into effective performance in the present and prepares capable leaders for the future. The Air Force needs these leaders to accomplish the national objectives set for national security to defend the safety of our people and nation when those objectives require the use of armed force.

Leadership does not equal command, but all commanders should be leaders. Any Air Force member can be a leader and can positively influence those around him or her to accomplish the mission. This is the Air Force concept of leadership, and all aspects of Air Force leadership should support it. The vast majority of Air Force leaders are not commanders. These individuals, who have stepped forward to lead others in accomplishing the mission, **simultaneously serve as both leaders and followers at every level of the Air Force**, from young Airmen working in the life support shop, to captains at wing staffs, to civilians in supply agencies, to generals at the Pentagon. Desirable behavioral patterns of these leaders and followers are identified in this doctrine and should be emulated in ways that improve the performance of individuals and units. Leaders positively influence their entire organization, without necessarily being the commander.

Mission

The primary task of a military organization is to perform its mission. The leader's primary responsibility is to motivate and direct people to carry out the unit's mission successfully. A leader must never forget the importance of the personnel themselves to that mission.

People

People perform the mission. They are the heart of the organization and without their actions a unit will fail to achieve its objectives. A leader's responsibilities include the care, support, and development of the unit's personnel. Successful leaders have continually ensured that the needs of the people in their unit are met promptly and properly.

An **Airman** is any US Air Force member (officer or enlisted; active, reserve, or guard; and Department of the Air Force civilians) who supports and defends the US Constitution and serves our country. An Airman understands the potential of air and space power.

All Air Force leaders share the same goal, to accomplish their organization's mission. Upon entering the Air Force, members take an oath, signifying their personal commitment to support and defend the Constitution of the United States and a commitment and willingness to serve their country for the duration of their Air Force career. The oath is a solemn promise to do one's duty and meet one's responsibilities. The oath espouses the responsibility to lead others in the exercise of one's duty. (The oaths of office and enlistment for officers, civilians, and enlisted personnel are at Appendix A.)

Regardless of duty location, occupational specialty, or job position, all Airmen must embody the **warrior ethos**, tough-mindedness, tireless motivation, an unceasing vigilance, and a willingness by the military members to sacrifice their own lives for their country if necessary. Air Force Airmen, military and civilian, are committed to being the world's premier air and space force. This is the warrior ethos.

The abilities of a leader, which are derived from innate capabilities and built from experience, education, and training, can be improved upon through deliberate development. Using the Air Force leadership components as described below is the means by which Airmen can achieve excellence—by living the Air Force core values, developing enduring leadership competencies, acquiring professional and technical competence, and then acting on such abilities to accomplish the unit's mission, while taking care of the unit's personnel. Core values permeate leadership at all levels, at all times. Leaders at the more junior levels must demonstrate the personal leadership competencies needed to create a cohesive unit fully supportive of its mission. Mid-level leaders will use the people/team leadership competencies to advance the organization's responsibilities within the framework of the operational mission. The more senior the leader, the more crucial becomes the influence on the institutional excellence of the organization. The ability to influence people, improve performance, and accomplish a mission-- the leadership actions--is part of all levels of leadership.

Leadership is the art of getting someone else to do something you want done because he wants to do it.

—**President Dwight D. Eisenhower**

President (then General) Eisenhower addressing troops before D-Day

Safety and risk management are crucial to effective operations through the preservation of Air Force people, resources and capabilities. As leaders, commanders at all levels should implement effective hazard identification, risk management and mitigation processes. Safety is a key leadership responsibility that cannot be relinquished or delegated. As it is people who are responsible for mission accomplishment, their safety is vital to that mission.

Risk is inherent in military operations. It is incumbent on commanders to avoid unnecessary risk to conserve lives and resources. The most efficient means to accomplish this is through effective risk management. This does not inhibit a commander's flexibility and initiative to perform assigned missions. While risk management does not remove risk entirely, it does allow a commander to control and minimize risk to an acceptable level.

Commanders are ultimately responsible for the management of risk and for safety processes, yet it is incumbent on all Airmen to make safety an intrinsic element of their activities, both on and off duty. In so doing, they play their role in supporting the leader's role in keeping the dual parts of 'people' and 'mission' functioning in the best manner possible.

LEADERSHIP COMPONENTS

In the Air Force, leadership contains three main components: core values, competencies, and actions. Leaders apply these components at three leadership echelons: the tactical level, the operational level, and the strategic level.

✪ **Air Force Core Values.** These are the values the Air Force demands of all its members at all levels. They are the guiding characteristics for all Air Force leaders.

○ **Leadership Competencies.** These are the occupational skill sets and enduring leadership competencies that Air Force leaders develop as they progress along levels of increased responsibility.

We have fulfilled the dreams of the visionary leaders who founded the Air Force, creating decisive and compelling effects from high above. We have renewed our focus on joint operations and the importance of integration with ground forces. Generals Arnold and Quesada would be proud. And we have demonstrated to the world the professionalism, competence, and incredible skill of Airmen -- men and women steeped in the warrior ethos and prepared to sacrifice their lives in the service of a cause greater than themselves.

—James G. Roche
Secretary of the Air Force, 2003

○ **Leadership Actions.** These are the actions leaders use to get things done. Air Force leaders influence and improve their units in order to accomplish their military **mission.**

Air Force Core Values

The Air Force core values are the bedrock of leadership in the Air Force. **The core values are a statement of those institutional values and principles of conduct that provide the moral framework within which military activities take place. The professional Air Force ethic consists of three fundamental and enduring values of integrity, service, and excellence.** This ethic is the set of values that guide the way Air Force members live and perform. Success hinges on the incorporation of these values. In today's time-compressed, dynamic, and dangerous modern battlespace an Airman does not have the luxury of examining each issue at leisure. He or she must fully internalize these values so as to know how to automatically act in all situations—to maintain integrity, to serve others before self, and to perform with excellence and encourage it in others.

Integrity First

Integrity is the fundamental premise for military service in a free society. Without integrity, the moral pillars of our military strength, public trust, and self-respect are lost.

—General Charles A. Gabriel
CSAF, 1982-1986

Integrity is the willingness to do what is right even when no one is looking. It is the "moral compass"—the inner voice, the voice of self-control, the basis for the trust imperative in today's Air Force.

Integrity is the single most important part of character. It makes Airmen who they are and what they stand for as much a part of their professional reputation as their ability to fly or fix jets, run the computer network, repair the runway, or defend the airbase. Airmen must be professional, both in and out of uniform. Integrity is not a suit that can be taken off at night or on the weekend or worn only when it is important to look good. Instead, it is the time that we least expect to be tested when possessing integrity is critical. People are watching us, not to see us fail, but to see us live up to their expectations of us. Anything less risks putting the heritage and reputation of the Air Force in peril.

Integrity is the ability to hold together and properly regulate all the elements of one's personality. A person of integrity acts on conviction, demonstrating impeccable self-control without acting rashly.

Integrity encompasses many characteristics indispensable to Airmen:

- ✪ **Courage.** A person of integrity possesses moral courage and does what is right even if the personal cost is high.
- ✪ **Honesty.** In the Service, one's word is binding. Honesty is the foundation of trust and the hallmark of the profession of arms.
- ✪ **Responsibility.** Airmen acknowledge their duties and take responsibility for their own successes or failures. A person with integrity accepts the consequences of actions taken, never accepting or seeking undue credit for the accomplishments of others.
- ✪ **Accountability.** No Airman with integrity tries to shift the blame to others; "the buck stops here" says it best.
- ✪ **Justice.** Airmen treat all people fairly with equal respect, regardless of gender, race, ethnicity, or religion. They always act with the certain knowledge that all people possess fundamental worth as human beings.
- ✪ **Openness.** As professionals, Airmen encourage a free flow of information within the organization and seek feedback from superiors, peers, and subordinates. They never shy from criticism, but actively seek constructive feedback. They value candor in their dealings with superiors as a mark of loyalty, even when offering dissenting opinions or bearing bad news.
- ✪ **Self-respect.** Airmen respect themselves as professionals and as human beings. Airmen with integrity always behave in a manner that brings credit upon themselves, their organization, and the profession of arms.
- ✪ **Humility.** Airmen comprehend and are sobered by the awesome task of defending the Constitution of the United States of America.
- ✪ **Honor.** All Airmen function in their Service with the highest traditions of honoring the Air Force's responsibilities to the nation and the sacrifices of its predecessors. It is incumbent on Airmen to uphold these traditions, adhering to what is right, noble, and magnanimous.

Service Before Self

As an Air Force core value, service before self represents an abiding dedication to the age-old military virtue of selfless dedication to duty at all times and in all circumstances—including putting one's life at risk if called to do so. The service-before-self value deals with accepting expeditionary deployments and assignments, accomplishing a job without scheming to accept jobs that get "face time" while others have to do the mission. Further, service before self does not mean service before family. Airmen have a duty to their family as strong as that to the Service. The difference is, there are times the Service and nation will require them to be away from home. Their responsibilities to their family include preparing and providing for them when deployed or when duty away requires it. The moral attributes stemming from this core value include:

In August 1950, while working at the base bakery at what is now Travis Air Force Base, CA, Sergeant Paul Ramoneda witnessed a B-29 crash in front of his duty station. Sergeant Ramoneda and three fellow Airmen ran toward the aircraft and rescued eight crewmembers. Sergeant Ramoneda returned to the plane determined to rescue more of the crew. The subsequent explosion killed Sergeant Ramoneda and the remainder of the crew, leaving a 60-foot crater. His selfless act was in the finest tradition of the Air Force core value of Service Before Self.

✪ **Duty.** Airmen have a duty to fulfill the unit's mission. Service before self includes performing to the best of one's abilities the assigned responsibilities and tasks without worrying how a career will be affected. Professionals exercise judgment while performing their duties; they understand rules exist for good reason. They follow rules unless there is a clear operational or legal reason to refuse or deviate.

✪ **Respect for Others.** Good leaders place the welfare of their peers and subordinates ahead of personal needs or comfort. Service professionals always act in the certain knowledge that all people possess a fundamental worth as human beings. Tact is an element of this respect.

✪ **Self-discipline.** Air Force leaders are expected to act with confidence, determination, and self-control in all they do in order to improve themselves and their contribution to the Air Force mission. Professionals refrain from openly displaying self-pity, discouragement, anger, frustration, or defeatism.

✪ **Self-control.** Service professionals—and especially commanders at all echelons—are expected to refrain from displays of anger that would bring discredit upon themselves and the Air Force. Leaders are expected to exercise control in the areas of anger, inappropriate actions or desires, and intolerance.

> *Anyone can become angry—that is easy. But to be angry with the right person, to the right degree, at the right time, for the right purpose, and in the right way—that is not easy.*
>
> **—Aristotle**

✪ **Appropriate Actions or Desires.** Leaders are guided by a deeply held sense of honor, not one of personal comfort or uncontrolled selfish appetites. Abuse of alcohol or drugs, sexual impropriety, or other undisciplined behavior is incompatible with military service. It discredits the profession of arms and undermines the trust of the American people. All Airmen maintain proper professional relationships with subordinates, superiors, and peers.

✪ **Tolerance.** Leaders understand an organization can achieve excellence when all members are encouraged to excel in a cooperative atmosphere free from fear, unlawful discrimination, sexual harassment, intimidation, or unfair treatment.

✪ **Loyalty.** Airmen should be loyal to their leaders, fellow Airmen and the institution they serve. American military professionals demonstrate allegiance to the Constitution and loyalty to the military chain of command and to the President and Secretary of Defense, regardless of political affiliation.

Excellence in All We Do

> *The power of excellence is overwhelming. It is always in demand and nobody cares about its color.*
>
> **—General Daniel S. "Chappie" James**
> **Commander in Chief, North American Aerospace**
> **Defense Command, 1975-1977**

This core value demands Airmen constantly strive to perform at their best. They should always strive to exceed standards objectively based on mission needs. This demands a continuous search for new and innovative ways of accomplishing the mission. There are several aspects of excellence: personal, organizational, resource, and operational.

✪ **Personal Excellence.** Airmen seek out and complete developmental education, stay in top physical, mental, and moral shape, and continue to refresh their professional competencies. Airmen must ensure their job skills, knowledge, and personal readiness are always at their peak.

✪ **Organizational Excellence.** Organizational excellence is achieved when its members work together to successfully reach a common goal in an atmosphere that preserves individual self-worth. No Airman wins the fight alone—even the single-seat fighter pilot relies upon scores of maintenance and support personnel to accomplish every sortie. Leaders foster a culture that emphasizes a team mentality while maintaining high standards and accomplishing the mission.

☣ **Resource Excellence.** Understanding that budgets are not limitless, Air Force leaders aggressively protect and manage both human and material resources. The most precious resource is people, and an effective leader does everything to ensure all personnel are trained, fit, focused, and ready to accomplish their missions. Leaders effectively use their resources to perform assigned tasks and understand they should only obtain resources necessary to accomplish their missions.

☣ **Operational Excellence.** The Air Force leader understands that all efforts in developing and employing air and space forces are directed at providing unmatched air and space power to secure the national interests of the United States. Airmen should prepare for joint and multinational operations by learning the doctrine, capabilities, and procedures of other US Services and allied forces.

The Air Force recognizes these **core values as universal and unchanging** in the profession of arms. They provide the standards with which to evaluate the ethical climate of all Air Force organizations. Finally, when needed in the cauldron of war, they are the beacons vectoring the individual along the path of professional conduct and the highest ideals of integrity, service, and excellence.

Leadership Competencies

Each of the three leadership levels within the Air Force is distinct from but related to the levels of warfare and requires a different mix of competencies and experience. Leadership at the tactical level is predominantly direct and face-to-face. As leaders ascend the organizational ladder to the operational level, leadership tasks become more complex and sophisticated. Strategic leaders have responsibility for large organizations or systems. The construct model for leadership, shown in figure 1.1, portrays how the leadership competencies will vary in their degree of use based on which leadership level is in use. While all aspects of the competencies are necessary to varying degrees at all levels, there is a change in focus based on the level at which a leader is operating.

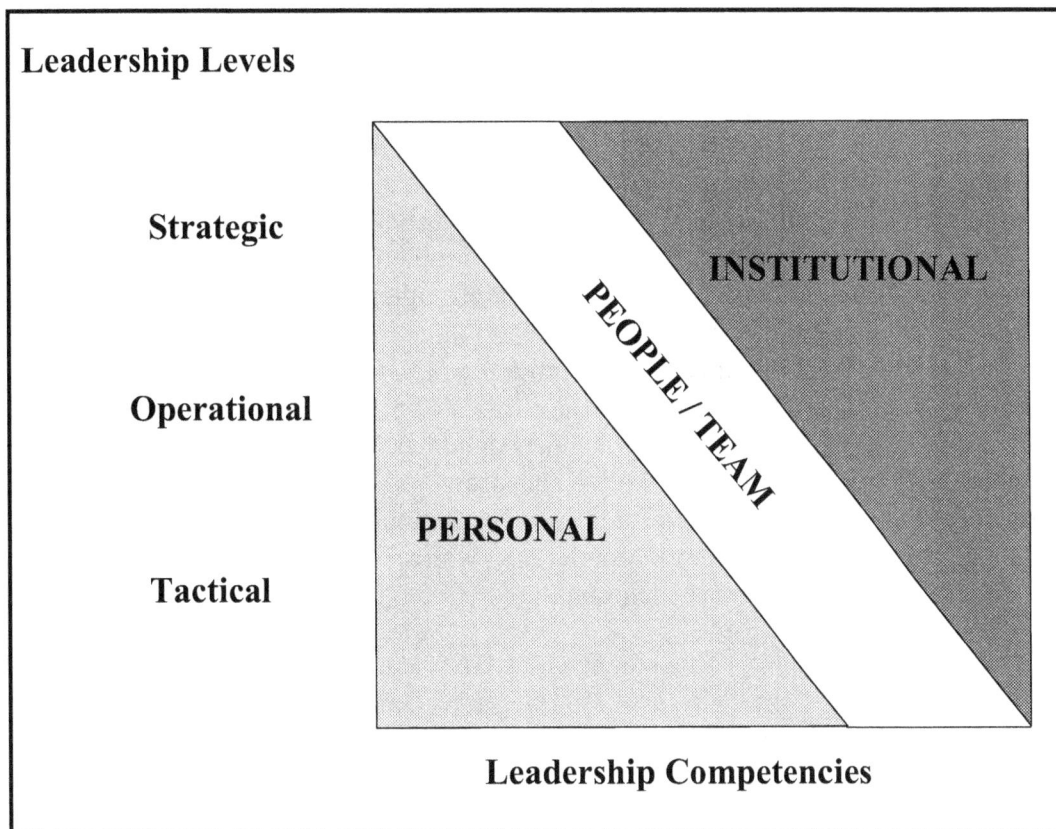

Figure 1.1. Relationship of Leadership Levels with Enduring Leadership Competencies

As leaders move into the most complex and highest levels of the Air Force, or become involved in the strategic arena, the ability to conceptualize and integrate becomes increasingly important. Leaders at this level focus on establishing the fundamental conditions for operations to deter wars, fight wars, or conduct operations other than war. They also create organizational structures needed to deal with future requirements. Figure 1.1 provides a graphic representation of the flow of leadership competencies through the levels of leadership; the accompanying text below then elaborates on the specifics of these relationships.

✪ **Personal Leadership:** This competency focuses on face-to-face, interpersonal relations that directly influence human behavior and values. Personal leadership is exercised at all levels—tactical, operational, and strategic. It is required to build cohesive units and to empower immediate subordinates. Skill sets required at the tactical level include knowing the technical and tactical competence of individual Airmen and having interpersonal skills, in addition to performing leader tasks, problem solving, performance counseling, and followership that implements policies and accomplishes missions. Personal leadership at the tactical level focuses on short range planning and mission accomplishment. At the tactical level of leadership, a Security Forces fire team leader should focus on the needs and abilities of the personnel he or she is tasked to supervise; this should be his or her main focus. At the operational level a master sergeant at the major command (MAJCOM) or numbered air force (NAF) level should have superior interpersonal skills to lead the enlisted force. At the strategic level. MAJCOM or

Headquarters USAF, senior officials apply their expertise in developing plans and programs to guide the Air Force toward achieving the Air Force mission.

- ✪ **People/Team Leadership:** This competency involves more interpersonal relations and team relationships. Leaders using this competency tailor resources to organizations and programs and, when in command, set command climate. Skill sets required for effective people/team leadership include technical and tactical competence on synchronizing systems and organizations, sophisticated problem solving, interpersonal skills (emphasizing listening, reading, and influencing others indirectly through writing and speaking), shaping organizational structure and directing operations of complex systems, tailoring resources to organizations or programs, and establishing policies that foster a healthy command climate. These leaders focus on mid-range planning and mission accomplishment. A squadron commander should be able to focus simultaneously on teambuilding, creating a more effective squadron, yet still be ready to handle individual personnel issues that arise. An air and space operations center director has the challenge of leading personnel in intense operations, while monitoring the conditions of the personnel themselves to ensure their continued effectiveness.

- ✪ **Institutional Leadership:** This competency exists at all levels throughout the Air Force, predominantly at the strategic level. Strategic leaders apply institutional leadership to establish structure, allocate resources, and articulate strategic vision. Effective institutional leadership skill sets include technical competence on force structure and integration; on unified, joint, multinational, and interagency operations; on resource allocation; and on management of complex systems; in addition to conceptual competence in creating policy and vision and interpersonal skills emphasizing consensus building and influencing peers and other policy makers—both internal and external to the organization. This level is the nexus of warfighting leadership skills for the Air Force. It is achieved through having learned the lessons from the earlier leadership competencies. For example, commanders, chief master sergeants, and civilian directors use institutional leadership to integrate people and teams, performing diverse tasks, to achieve mission accomplishment. A Commander, Air Force Forces (COMAFFOR) must be able to operate at this level in the rapidly changing environments of combat and other operations where the Air Force functions.

Leadership skills needed at successively higher echelons in the Air Force build on those learned at previous levels. As military and civilian leaders progress within the Air Force, they serve in more complex and interdependent organizations, have increased personal responsibility and authority, and have significantly different occupational competencies and enduring leadership competencies than their subordinates.

Enduring Leadership Competencies

Enduring leadership competencies are the personal and leadership qualities that should be common to all Air Force members. These enduring competencies are listed in figure 1.2, and they are defined at Appendix B. Occupational skill sets are derived from the enduring leadership competencies and inculcated by Airmen in their activities. (See Chapter Two.)

By deliberately exposing people to a broader range of experiences, by ensuring each Airman's developmental experience is both valuable and meaningful, and by cultivating the enduring leadership competencies, the Air Force creates leaders who are more flexible and adaptable in a force that has an even greater sense of belonging and importance.

Leadership Actions

Air Force leaders **influence** their subordinates through tools that include communication, motivation, standards, and decisiveness. The result is a unit moved to effectively perform a mission. Air Force leaders also **improve** their unit's abilities through development and learning. The result is an enhanced ability to **accomplish** the unit's assigned missions. Leaders influence and improve their units in order to accomplish their military mission.

✪ **Influence.** Leaders motivate and inspire people by creating a vision in their mind of a desirable end-state and keeping them moving in the right direction to achieve that vision. To do this, leaders tailor their behavior toward their fellow Airmen's need for motivation, achievement, sense of belonging, recognition, self-esteem, and control over their lives.

PERSONAL LEADERSHIP
- Exercise Sound Judgment
- Adapt and Perform Under Pressure
- Inspire Trust
- Lead Courageously
- Assess Self
- Foster Effective Communication

LEADING PEOPLE/TEAMS
- Drive Performance through Shared Vision, Values, and Accountability
- Influence through Win/Win Solutions
- Mentor and Coach for Growth and Success
- Promote Collaboration and Teamwork
- Partner to Maximize Results

LEADING THE INSTITUTION
- Shape Air Force Strategy and Direction
- Command Organizational and Mission Success through Enterprise Integration and Resource Stewardship
- Embrace Change and Transformation
- Drive Execution
- Attract, Retain, and Develop Talent

Figure 1.2. Enduring Leadership Competencies

President George W. Bush at Thanksgiving 2003 dinner in Iraq

They (your troops) want to know how much you care, long before they care how much you know.
—Lieutenant General Robert Springer, USAF USAF Inspector General, 1985–1988

✪ **Improve.** Leaders foster growth by insisting that their people focus attention on the aspects of a situation, mission, or project they control. Challenge should be an integral part of every job; for people to learn and excel, they must be intrinsically motivated. Leaders should provide challenging and enlightening experiences. It is important to identify and analyze success in order to make the causes and behaviors permanent and pervasive, not temporary and specific. Leaders encourage the learning process by formally recognizing individual and unit success, no matter how large or small. Leaders create more leaders.

✪ **Accomplish.** Air Force leaders influence people, improve their abilities, and direct their activities to accomplish their military mission. Leaders produce the effects that successfully achieve desired objectives.

I think it's important that leadership opportunities be forced to as low a commissioned level as possible. That's a good time to start looking at a guy. For heaven's sake! What can you put in his ER besides the fact that the guy is a great stick, he does well on instruments, is a good gunner, and he talked to the Kiwanis club once a quarter downtown. What else can you put in there? How do you separate the wheat from the chaff? Get him in a leadership role!

—General Hoyt S. Vandenberg
CSAF, 1948-1953

CHAPTER TWO

FORCE DEVELOPMENT IN THE AIR FORCE

> *Force Development will enable us to focus on each individual by emphasizing our common Airman culture.... Every aspect of Force Development has one common goal: To continue developing professional Airmen who instinctively leverage their respective strengths together. We intend to develop leaders who motivate teams, mentor subordinates, and train successors.*
>
> **—General John P. Jumper, CSAF, 2002**

People are the Air Force's most critical asset: The Air Force's Airmen—its active duty, guard, and reserve component officers, enlisted, and government civilian workforce—turn tools, tactics, techniques and procedures into power projection, battlespace effects, and global mobility. For this reason, the art of employing Airmen with the requisite skills, knowledge, experience, and motivation is a foundational endeavor, with effects on current operations and future capabilities.

The key roles Airmen play in planning, preparing, organizing, and executing all the elements necessary to conduct Air Force operations are critical to mission success. Even with the most sophisticated aircraft and technology available, the ability of people to apply those tools determines the effectiveness of the force. Accordingly, the Air Force invests time and resources in its Airmen to develop their abilities and train them in the desired skills. The importance of Airmen to Air Force capabilities and the effects they achieve in pursuit of US objectives demands that the Air Force maximize the competencies possessed by those Airmen and effectively integrate them with technology, strategy, and tactics to produce the right operational capabilities and effects.

Developing Airmen best happens through a deliberate process, one that aims to produce the right capabilities to meet the Air Force's operational needs. It focuses on developing leaders who thoroughly understand the mission, the organization, and the tenets of air and space power. Deliberate force development is a systematic process that gives present and future leaders a broad perspective of Air Force capabilities, while simultaneously developing individual occupational skills and enduring competencies. These two complementary tracks of force development are essential components of tailoring the right development to the right person at the right time. By increasing and developing enduring leadership competencies as people assume greater responsibility in the organization, the Air Force fosters leaders who understand the full spectrum of operations to accomplish missions across the entire organization.

The Air Force has been successful in producing many capable leaders. However, the evolution of technology and world political events demands a leadership development approach that keeps pace. The Air Force requires a simple, understandable, deliberate development system for the warfighter that:

✪ Satisfies necessary skill and enduring leadership competency needs,

✪ Is institutionalized by doctrine,

✪ Is driven by policies and connected resources that concentrate on the right level and focus on experience, education, and training, at the right time, and

✪ Optimizes the amount of time available for force development.

A phased approach at tactical, operational, and strategic levels provides the framework for focusing development of occupational skill sets and enduring leadership competencies. **Force development is a series of experiences and challenges, combined with education and training opportunities that are directed at producing Airmen who possess the requisite skills, knowledge, experience, and motivation to lead and execute the full spectrum of Air Force missions.** It is the method the Air Force uses to grow experienced, inspirational leaders who have the necessary technical competencies and professional values, framed by a common culture, regardless of career specialty. Force development is the essence of capability to accomplish the Air Force mission.

Early in an Airman's career, development is aimed primarily at personal competencies at the tactical level with an introduction to people/team leadership. At the operational stage, personal leadership continues, but much greater emphasis falls into people/team leadership development, and institutional leadership competencies are introduced. At the strategic level, the greatest emphasis is on developing institutional leadership competencies in preparing Airmen to be senior leaders and commanders.

FORCE DEVELOPMENT CONSTRUCT

The need for Airmen, military and civilian, who possess the right occupational skill sets and enduring leadership competencies forms the core of force development and is the basis for all force development efforts. The goal of force development is to prepare Airmen to successfully lead and act in the midst of rapidly evolving environments, while meeting their personal and professional expectations. The construct, as depicted in figure 2.1, starts with understanding mission requirements and translating them into capabilities. Doctrine takes those requirements and translates them into best practices for the Service. It establishes the bedrock capabilities of the Air Force that it brings to all joint operations, within which force development is then used to create leaders and commanders. Doctrine guides the presentation and employment of Air Force capabilities.

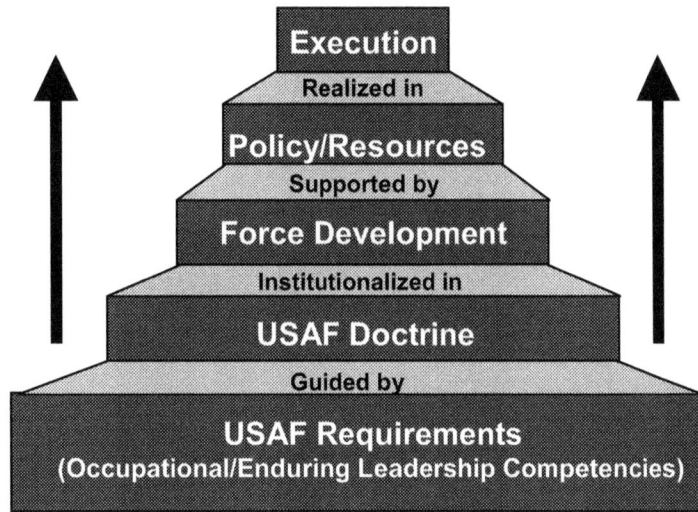

Figure 2.1. Force Development Construct

These capabilities are then refined to determine which competencies are required to accomplish the associated tasks, as well as the workload involved. In this process, occupational skill sets needed to do the job are taught in technical training or learned on the job. Force development recognizes the need for a deep perspective in a functional area or occupational skill, but at the same time offers the means to achieve the wider perspective needed for leadership. Occupational skill sets are driven by position requirements and promoted by systematic, deliberate development. Force development programs specify how the Air Force leverages its investment in its people. The Air Force has determined there are clearly identifiable skill requirements for Airmen who have experiences in more than one connected career area. Force development defines the occupational skill combinations and then facilitates the education, training, and assignment processes to produce a sufficient capability within the personnel inventory. Some possible occupational skill combinations for Air Force officers are illustrated in figure 2.2 below:

Notional Officer Occupational Skill Combinations:

Core Skill → **Enhanced With** →	**Development Skill**
Airlift Pilot/Navigator	Acquisition Management
Acquisition Management	Plans and Programs
Space	Plans and Programs
Fighter Pilot	Space Operations

Figure 2.2. Notional Officer Occupational Skill Combinations

Force development is executed through policies, force management strategies, and prioritization of resources. Finally, these programmatic decisions are executed through deliberate management of Air Force programs and operations in the field to achieve the desired objectives.

LEVELS OF FORCE DEVELOPMENT

Force development is based on producing desired effects through Airmen's professional and technical development. The key to this result is understanding what is "right" in each element of this broad objective and having the ability to achieve synergy through the complex system of interactions that contribute to the desired capabilities. These interactions also exist at the tactical, operational, and strategic levels.

Tactical Level

At the tactical or the personal/direct level, Airmen master their primary duty skills. They also develop experiences in applying those skills and begin to acquire the knowledge and experience that will produce the competencies essential to effective leadership. Tactical leaders are the Air Force's technicians and specialists. **At this level, Airmen learn about themselves as leaders and about how their leadership acumen can affect others.** They are focused on honing followership abilities, influencing peers and motivating subordinates. They are learning about themselves and their impact on others in roles as both follower and leader. They are being assimilated into the Air Fore culture and are adopting the core values of their profession. They are gaining a general understanding of team leadership and an appreciation for institutional leadership. The tactical level of the Air Force encompasses the unit and sub-unit levels where individuals perform specific tasks that, in the aggregate, contribute to the execution of operations at the operational level. Tactical level performance embraces flying an aircraft, guarding a perimeter, loading a pallet, setting up a firewall for a base, identifying a potentially hostile radar return, treating a broken arm, and many other forms of activity, accomplished by both military and civilian personnel.

Operational Level

At the operational or organizational level, Airmen are able to understand the broader Air Force perspective and the integration of diverse people and their capabilities to execute operations. This level is where an Air Force member transitions from being a specialist to understanding Air Force integration. Based on a thorough understanding of themselves as leaders and followers and how they influence others, they apply an understanding of organizational and team dynamics. They continue to develop personal leadership skills, while developing familiarity in institutional leadership competencies. The operational level includes continued broadening of experience and increased responsibility within a family of complementary skills. This will normally follow intermediate developmental education and extends beyond simple career broadening; it deliberately focuses developmental activity to produce the right skills mix at the appropriate level to meet Air Force requirements. Assignment to squadron command and similar positions of authority, such as division/branch chief, occur during this phase of development. **The focus of Air Force organization and employment is at the operational level. It is here where warfighting is executed and the day-to-day command and control of Air Force operations are carried out.** At this level the tactical skills and expertise Airmen developed earlier are employed alongside new leadership opportunities to affect an entire theater or joint operations area. By now the Airman, military and civilian, has developed a family of related skills, grounded in the Air Force expeditionary culture and enduring leadership competencies, guided by ingrained core values.

At a warfighting numbered air force (NAF), major command (MAJCOM) staff, or theater air and space operations center, the operational competence of the Airman is tested and grown. Those Airmen with experience in more than one career area have an advantage over others in preparing them for future leadership roles requiring even broader perspectives. This perspective broadening should follow a logical and deliberate path to link experience in related areas and to fulfill the operational needs of the Air Force. This linkage should parallel the relationship that exists among various air and space capabilities within a specialty (for example, an Airman involved in acquisition should be experienced not only in acquisitions management but also in contracting, air and space power employment, plans and programs, and similar specialties). This method represents a "packaged" approach to competency development. By deliberately exposing Airmen to a broader range of experiences for which they are prepared, the experience becomes more meaningful.

Strategic Level

At the strategic level, Airmen combine highly developed occupational and enduring competencies to apply broad professional leadership capabilities. They develop a deep understanding of Air Force missions and how operational capabilities and Airmen are integrated to achieve synergistic results and desired effects. They also understand how the Air Force operates within joint, multinational, and interagency relationships. At this level, an Airman's required competencies transition from the integration of people with missions, to leading and

directing exceptionally complex and multi-tiered organizations. Based on a thorough understanding of themselves as leaders and followers, and how to use organizational and team dynamics, they apply an in-depth understanding of leadership at the institutional and interagency levels. They achieve a highly developed, insightful understanding of personal and team leadership, while mastering their institutional leadership competencies. The strategic level includes challenges to gain breadth of experience and leadership

As Defense Logistics Agency Vice Director, Major General Mary L. Saunders operates at the strategic level of force development.

perspective (e.g., logical pairings of skills; educational opportunities, and training focused on the institutional Air Force; joint, intergovernment, business and international views). The strategic level of development focuses on the effects an Airman, military and civilian, can have across a MAJCOM, a theater, the Air Force, or even other Services or the Department of Defense. The seeds of leadership that were planted during tactical skills development and matured into operational-level capabilities should bear fruit at the strategic level. **Senior leaders need tactical comprehension and competence, as well as broader perspectives and the ability to effectively lead Airmen and joint forces in an expeditionary environment.** They should embody Air Force cultural and core values that were nourished throughout the individual's career. At the strategic level of development Airmen receive further opportunities to expand their breadth of experience and have the greatest ability to influence the Air Force's role in military operations.

CHAPTER THREE

FORCE DEVELOPMENT: THE PROCESS

MAXIMIZING CAPABILITY

Force development processes are focused to produce and maximize the capabilities of Airmen. Meeting the need for trained, knowledgeable, experienced, and motivated Airmen starts long before a person enters the Air Force, and transcends a single commander's need for a specified capability. The elements of force development are deliberately managed to provide a full spectrum of human competencies where and when needed. The continuous expansion of capabilities possessed by each Airman and the overall force is achieved through deliberate developmental activity. The needs of each Airman remain an important element in developing and sustaining a ready and willing force. Force development aims to optimize the capabilities of the individual, the unit, and the Air Force, while balancing personal needs with mission mandates. When force development shifts its primary focus to tactical processes or individual aims, Airmen can be erroneously viewed as commodities, and the whole force suffers.

A focus on the four elements of the force development process—**force development definition, renewal, development, and sustainment**—supported by proper planning, will produce the balanced, diverse, and capable workforce the nation requires. These four elements combine to produce effects greater than the sum of their parts. This process is depicted in figure 3.1.

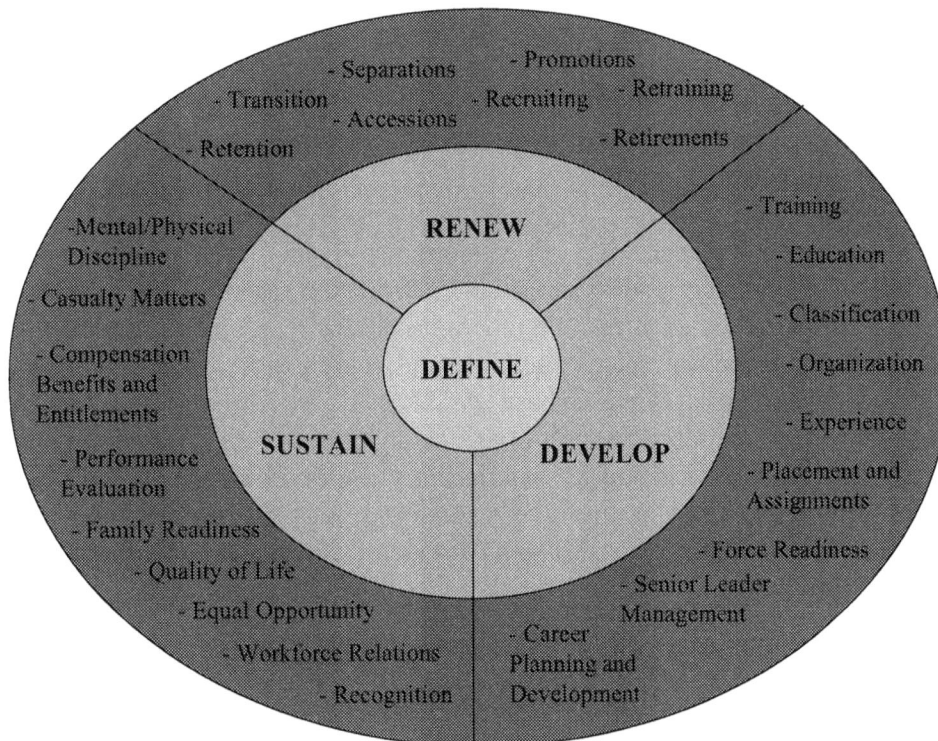

Figure 3.1. Force Development Process

Force Development Process Definition

Effective integration of force development is based on capability-based requirements. Success of Air Force operations depends on the effective integration of human capabilities with the tools, tactics, techniques, and procedures that combine to produce the full spectrum of air and space power. The first steps in integrating people into Air Force operations are defining the required capabilities, then organizing the skill sets required to produce those capabilities. The expeditionary Air Force has transformed from a forward deployed, in-garrison force trained for one primary mission and adversary into a flexible expeditionary force, capable of responding to a number of mission scenarios. Therefore, our leadership and force capabilities must be developed to meet Air Force expeditionary requirements.

Renewal

Force renewal ensures the Air Force continues, over time, to maintain the capability of its workforce to execute its mission. As human requirements and capabilities of the workforce change over time, the processes associated with renewal are essential to the continued readiness of the force. Renewal seeks to maintain the right balance across the force. As many dynamics of military service demand vigor, the renewal process seeks to balance those attributes with the wisdom of age and experience required for leadership and technical mastery. The renewal of force capabilities is also a key dynamic in determining the right mix of active duty, reserve, and civilian elements. An effective mix and balance of individual characteristics are essential to developing occupational skills and leadership competencies that satisfy Air Force mission requirements.

The Air Force renews its force to preserve and increase the capabilities of Airmen. Because workforce capabilities are highly sensitive to political, economic, and environmental factors, both internal and external, they can be perishable. Due to the effects of combat operations and other circumstances, such as economic shifts and force structure changes, the Air Force has a continuing requirement to renew and refresh the essential characteristics possessed by the force. This renewal is based on operational doctrine and is achieved by pursuing deliberate policies that drive accession, retention, and competency redistribution, in the form of retraining, based on Air Force needs. These elements are synchronized to acquire and retain necessary competencies, and divest capabilities no longer required. Advertising and recruiting strategies, Air Force precommissioning, basic military training objectives, and pipeline training capabilities are crucial elements in the force renewal process, driving the Air Force's ability to attract, procure, and produce the right skills and competencies to meet operational requirements.

Promotion of Airmen recognizes potential and provides a leadership cadre. Air Force promotion systems are intended to recognize and identify Airmen who have demonstrated potential for increased responsibility and assumption of more difficult, broadly based duties. This creates a two-part effect. For the Airman, both military and civilian, pride in accomplishment, improved quality of life, and enhanced personal motivation flow from this recognition. They are provided the means to achieve increased opportunities. For the Air Force, promotion represents a keystone to succession planning and execution. Linkage of rank/grade requirements to organizational placement and advancement to higher ranks clearly separate those who are candidates to enter the next level of institutional leadership from those who remain at a lower level of professional or technical development. In certain career areas, under constrained selection policies, a newly promoted individual may not always have the requisite technical expertise or professional experience to immediately succeed to a higher ranked/graded position. Therefore, it is vital the Air Force leverage its force development tools and selection processes to ensure sufficient numbers of Airmen possessing the skills, knowledge, and experience to effectively lead in those higher ranked/graded positions are identified for advancement. As promotion recognizes leadership potential, it also serves to motivate Airmen to aspire to lead by providing them with the opportunity for greater responsibility, compensation, and prestige. Developing a leadership succession process that serves organizational needs and meets the expectations of motivated individuals drives the Air Force to pursue promotion strategies that encourage development of leadership aspiration and competencies, yet balance them with reasonable promotion expectations and equitable advancement opportunities. This is an investment in the Air Force future.

Simultaneous promotions for a mother and daughter provide a leadership heritage for the Air Force.

Development

Development processes and systems take individual capabilities and, through education, training, and experience, produce skilled, knowledgeable, and competent Airmen who can apply the best tools, techniques, and procedures to produce a required operational capability.

The Air Force educates and trains Airmen to improve the capabilities of the force. The Air Force seeks to access potential Airmen from every segment of American society. These new entrants seldom possess the skills and knowledge they need to effectively perform their duties. Military operations function in a unique environment that requires highly specialized skills, knowledge, and experience. Development is the process where raw aptitudes, attitudes, and abilities are supplemented and transformed through a combination of specialized training, education, and hands-on experience, to craft the individual competencies which encompass the human capability the Air Force must possess to execute its air and space missions. A progressive program of professional, technical, and academic development from the earliest stages of education and training prepares Airmen to comprehend and accomplish their assigned missions.

Force development is multidimensional. Experience, blended with skills, knowledge, and motivation, refines occupational skills and enduring leadership competencies possessed by each Airman, military and civilian. Through deliberate career planning and development, Airmen are assigned and employed to meet mission requirements in ways that also capitalize on the Air Force's investment in training and education. The Air Force leverages these competencies to expand the capabilities of the entire workforce. As Airmen, both military and civilian, rise in levels of responsibility, they are exposed to and challenged by broader aspects of the organization. **The Air Force prepares Airmen for leadership by optimizing experiences and skills to provide an effective understanding of the appropriate levels of the organization and by developing capabilities to meet those challenges.**

The Air Force employs Airmen to achieve military objectives. The objective of all force development efforts is synergy of diverse competencies—the point at which motivation, leadership, training, education, experience, and the full range of human and technological capabilities are brought to bear. Although development continues through employment, operational objectives become the focus for the integration of people and technology to produce air and space power. It is in employing Airmen where requirements and developmental efforts merge to achieve results.

Civilians directly contribute to Air Force operations.

Sustainment

Sustainment processes and programs aim to balance retention and invested capability with cost effectiveness and evolve to respond to changing operational requirements. These processes are structured to capitalize on investments in force development. At some point, all Airmen will depart military service. The investment in time, effort, and resources should focus on deliberate outcomes that foster effective mission accomplishment. "Quality of life" programs clearly reflect the high value the Air Force places on its people. More significantly, they highlight the Service's need to retain the right mix of people with occupational skills and enduring leadership competencies and associated capabilities. The highly specialized and technical nature of Air Force operations drives a long-term need for diversity in skills, knowledge, and experience.

The Air Force sustains its force to maintain a high state of readiness. Sustaining the force ensures that sufficient numbers of appropriately skilled, knowledgeable, and experienced Airmen are properly motivated and ready to execute the full spectrum of Air Force missions. As readiness is a perishable state, the Air Force continuously assesses and addresses individual and unit readiness issues.

The Air Force provides competitive pay and benefits to attract, motivate, and sustain the force. An effective compensation system will motivate quality, experienced Airmen to remain with the Air Force, as well as attract sufficient highly qualified and diverse new recruits and employees.

The Air Force supports families, recognizing their value in motivating and sustaining the capabilities of Airmen. Career decisions and associated effects on retention and force readiness increasingly involve the opinions and affairs of families. While the Air Force recruits individuals, it retains members with families where spouses and children play a large role in retention related decisions. By supporting families and ensuring they are able to enjoy a quality of life and opportunities comparable to those enjoyed by the citizens of the nation they protect, the Air Force makes a tangible commitment to Airmen that demonstrates recognition and value of their service. These family initiatives not only increase the commitment of Airmen to the Air Force, but also enlist the support of valuable and influential allies. The impact of family influence on morale and retention is distinct and significant.

High standards for military members reinforce the discipline required by combatants and in application of deadly force. Part of basic military training involves indoctrination to a level of rigor and discipline not commonly present in civilian life. These realities of military service involve experiences that include facing life and death combat decision making and leadership challenges in extremely harsh operational and environmental conditions. Instilling rigor and discipline extends well beyond basic training to daily life in uniform. It is reinforced at every level of command and becomes an inherent characteristic of each Airman's self and image, and a common bond across all members of the Air Force. Dress and appearance standards, physical fitness standards, and customs and courtesies are a few examples of how the Air Force instills discipline in Airmen and keeps them mentally and physically prepared to respond to the challenges and demands of warfighting.

Physical fitness and discipline work together to improve an Airman's abilities for the Air Force.

Recognition of Airmen honors their achievements and motivates others to aspire to excellence. With the high risks associated with preparing for and engaging in military operations, it is appropriate to honor those who have valiantly fought and voluntarily sacrificed

23

their personal safety to assist their comrades or to triumph over an adversary. Similarly, exceptional performance cannot be recognized through financial remuneration, but should be formally acknowledged and reinforced. Air Force recognition programs are designed to highlight excellence in leadership and technical skill, as well as extraordinary dedication in performance of assigned duties or service to the nation.

Sustaining capability is done over time. Sustaining force capability is a long-term endeavor. The Air Force uses various means such as retention initiatives and stop loss, to mitigate the loss of competencies and experience, ensuring the force has requisite capabilities to perform assigned missions. Changing operational requirements, age, career choices, physical ailments, failure to maintain standards, and similar factors create the need to continually sustain the capability of the force. Keys to this sustainment effort are motivational tools such as recognition, compensation, and benefits that require a focused investment of limited resources to achieve specific force shaping outcomes.

In addition to sustaining force capability through retention in the regular force, there are other elements of sustainment that reap benefits for the Air Force in the long term. For those personnel separating or retiring after service in the regular Air Force, follow-on opportunities allow them to provide continued vital support to the Air Force and its responsibilities:

- ✪ If Airmen have not yet served in either the Air National Guard or the Air Force Reserve (the air reserve components or ARC), they may obtain benefits by transferring into one of these. In this manner, Airmen can continue pursuing the goals they have set for individual service to the Air Force, and the Air Force retains vital trained, educated, and motivated Airmen in a status that benefits the Service and the individual simultaneously.

- ✪ For those Airmen who have a desire to remain connected to the Air Force and avail the Service of their knowledge and experience, the option of civil service provides a means to do so. Continuing to serve the Air Force in the status of a civilian, similar to an Airman's involvement with the ARC, benefits both the Air Force and the individual. The oath of office for an Air Force civil servant is identical to that taken by an officer, bringing with it a comparable responsibility of service to the nation. Airmen transitioning from the uniform to civilian clothing bring with them extensive experience that is valuable to the Air Force and the nation.

- ✪ The Air Force places great reliance on its partnership with the civilian business sector of our nation's economy. Many Airmen, upon completion of service, enter the civilian workforce as contractors, able to offer their experience and knowledge in continued support to the Air Force.

In all cases above, Airmen who elect to sustain a relationship with the Air Force after transitioning from active duty provide a wealth of benefits in the form of knowledge and experience. Keeping this experience base active in the ARC or various civilian sectors is a vital part of the nation's overall defense capability.

PLANNING AND INTEGRATION

Coordinated strategic planning guides force development activity to produce competencies defined by operational capabilities. Planning, programming, policy, and process management activities must be coordinated to produce trained, knowledgeable, experienced, and motivated Airmen. This function underpins all the activity involved in developing and integrating Airmen into Air Force operations. It does so by ensuring all of the elements are effectively synchronized to deliver the right people at the right place at the right time, as required by operational needs. It complies with the strategic vision, guidance, associated goals, and direction of the President and Secretary of Defense, the Department of Defense, and the Department of the Air Force. With a coordinated strategic plan, each element of force development has a synergistic effect. The end product is the capability to perform the mission and achieve desired effects in both the near and long term.

CHAPTER FOUR

FORCE DEVELOPMENT THROUGH EDUCATION AND TRAINING

> *I'm firmly convinced that leaders are not born; they're educated, trained, and made, as in every other profession. To ensure a strong, ready Air Force, we must always remain dedicated to this process.*
>
> **—General Curtis E. LeMay**
> **CSAF, 1961-1965**

Education and training are critical components of the force development construct. Education and training represent a large investment of resources and are the primary tools in developing Airmen. They are the key mechanisms, or forcing functions, in the development phase of the force development process. They apply to the development of Airmen, military and civilian, at all levels—tactical, operational, and strategic. Education and training are distinct but related force development activities. Education provides critical thinking skills, encouraging exploration into unknown areas and creative problem solving. Its greatest benefit comes in unknown situations or new challenges. Thus, education prepares the individual for unpredictable scenarios. Conversely, training is focused on a structured skill set, and the results of training performance should be consistent. Thus, training provides the individual with skill expertise. Education and training together provide the tools for developing Airmen. For a further discussion on the differences between education and training, see Appendix C.

The force development process provides a deliberate approach to help commanders, supervisors, and individuals select the right education and training programs. This focus occurs across the tactical, operational, and strategic levels in an Airman's development. Training is typically the emphasis at the tactical level, education assumes greater importance as Airmen progress from the operational level to the strategic level, but each level is relevant to developing Airmen throughout their careers.

COMMON GUIDING PRINCIPLES

Certain common principles guide education and training at all levels of tactical, operational, and strategic development:

✪ **Build skill set expertise**. Design training programs with the experience and current skill levels of the participants in mind. Tactical-level training should take into account the relative inexperience and low knowledge level of young Airmen. Operational-level programs should leverage the skills already developed to broaden the Airmen's perspectives and capabilities, and offer them formal and informal leadership opportunities to enhance their growth. Strategic training should continue in this vein, shifting away from functional expertise and looking more at leadership and assessment skills, joint and coalition integration, and policy formulation. Training objectives should seek to identify and maximize individual abilities and mission required skills. Although initial training programs are uniform and standardized, continuation training should be adaptable and flexible to increase effectiveness for individuals. Every training program should have established performance or competency requirements to measure success in building the required skill sets. It is necessary to maintain a balance between academic concepts, operational reality, and lessons of the past.

Sure, everyone wants to be an effective leader, whether it be in the Air Force or in the community. You can and will be if you identify your strengths, capitalize on them, and consciously strive to reduce and minimize the times you apply your style inappropriately.

—Chief Master Sergeant of the Air Force Robert D. Gaylor
1977-1979

✪ **Prepare for change.** Skills development must keep pace with the changing operational environments and resulting changing requirements. Skills training at all levels concentrates on the known and therefore is normally stable in nature. Specialty and continuing training programs, however, should be proactive to changes in mission demands. Ultimately, the practice of preparing for change will prepare Airmen for leadership roles in a dynamic world. Technological and process change is inevitable, leading to evolutions in the Air Force operational art which Airmen must understand early so they can be prepared to be the Air Force leaders of tomorrow. Educators should survey training methods outside of organizational bounds (other Services, government, and industry) to stay abreast of new training and education insights and best practices and adapt these methods to the programs for training and educating the force. They should ensure educational programs are relevant using operational feedback mechanisms such as lessons learned, hot-wash sessions, after action reports, intelligence summaries, and similar current operations tools. Without active feedback programs, educational programs can become stagnant and less effective in preparing individuals to meet challenges in the field.

✪ **Create depth of expertise.** Competence and credibility require depth of experience that provides a foundation for effective leadership. Depth is not gained overnight, but is an expertise honed over time. Skills and leadership development programs should provide the fundamentals that will be reinforced by on-the-job training and expeditionary field experience. Commanders and supervisors should know that demanding duty assignments normally prove more effective in developing depth of expertise than do assignments that lack sufficient challenge. Tactical-level education and training should concentrate on building depth of knowledge and experience in the primary skill and skill-related areas, including an understanding of Air Force culture and values. Operational-level education and training should build on early skills and deepen their understanding of the complete Air Force employment capability and its interface with joint and coalition partners. At the strategic-level, education and training polish the leadership and command skills of Airmen and deepen their joint and coalition warfare and policy-making skills.

✪ **Train to mission demands.** Educators and trainers must assiduously seek the current requirements of the operational and warfighting community and craft programs to meet those requirements. Training that meets mission needs leverages both training resources and duty experience to maximum effect. Mission needs may require functional specialists to train in other areas (for example, the personnel specialist may be tasked to work in mobility processing or the communications-computer specialist may augment security force entry control;). Skill expertise should prepare individuals for all tasks they are expected to perform to meet mission demands, and these needs may change rapidly with policy or international events. Airmen should have general expertise in mission areas and specific expertise within their career field as demanded by the mission.

✪ **Train like we fight.** The US military believes that success hinges upon practicing the profession of arms in the same manner it will be executed on the battlefield or during a contingency. If training or exercises do not reflect realism, the stress and challenges of actual conditions, then our Airmen will not be prepared when they are called upon to execute their mission. Training programs must be aligned with expected outcomes and provide realistic experience to greatly improve skill competency. Stress, unpredictability, fatigue, night operations, adverse weather, simulated equipment breakdowns, and chemical and biological contamination are examples of the challenges our men and women will face in the field and should be trained to overcome.

✪ **Make training and education available.** In today's expeditionary Air Force and high tempo world, the opportunities to train or receive education have become more limited. Education and training must be responsive to this shift. Commanders and supervisors should ensure they take every opportunity to get their people the training or education they need for advancement and meeting unit mission requirements. Educators and trainers should continue to expand the opportunities for deployed training through online or other reachback capabilities. Furthermore, education and training programs may have to become shorter or more flexible and adaptive to the expeditionary environment to ensure coverage of the material in a study environment lacking continuity. Education and training are more challenging to get while deployed, but should not be neglected for that reason. Similarly, during redeployment and reconstitution, strong emphasis on getting Airmen into training and education programs will bring long-term rewards.

✪ **Validate education and training through wargames and exercises.** Simulation through exercises and wargames can be very effective in terms of time, cost, and experience gained toward preparing Airmen for their wartime and contingency roles. Exercises and wargames are effective methods of building individual experience under controlled conditions. In addition to providing instruction, these also aid in evaluating performance and the effectiveness of other training and education programs. Exercises can significantly contribute to training objectives, while wargaming is typically more appropriate for education in which critical thinking objectives are important. Exercises are also important tools in developing individual skill sets along with organizational capabilities. For example, the ULCHI FOCUS LENS exercise provides an opportunity for commanders and staffs to focus on operational and strategic issues associated with general military operations on the Korean peninsula. Wargames offer additional tools for developing and evaluating competencies of individuals and organizations. In August 2002, the GLOBAL MOBILITY Wargame analyzed and experimented with how the Air Force and Defense Department global mobility and logistics systems must come into play before and during the fight to make sure all the necessary people and equipment are deployed to the right place at the right time to meet mission requirements.

By embracing the training of future generations as a key principle of leadership, we ensure our successors are trained by professionals who pass on their knowledge and experience.

—General John Jumper, CSAF, 2003

Basic Military Training graduates represent the future of the Air Force

EDUCATION AND TRAINING AT THE TACTICAL LEVEL

Education and training at the tactical level includes training in a primary skill and education in the fundamentals of leadership. A new Airman must be educated in the common Airman culture and must understand the enduring values that bond Airmen together. In addition, they should receive an understanding of, and gain expertise in, their unique specialty. This is accomplished by the following tactical education and training activities:

✪ Fundamental education

✪ Specialty training

✪ Continuation training

✪ Leadership education

The following example illustrates use of the above activities: Junior enlisted Airmen, will complete basic training for indoctrination into Air Force culture, attend the relevant technical schools to obtain the occupational skills needed for their duties, and then receive orientation into their new organization at the local First Term Airman Center. As they gain experience and advance within their units, they will attend Airman Leadership School to enhance their ability to function as leaders within their organizations. Similarly, recently commissioned officers will obtain indoctrination to the Air Force through their commissioning sources, then receive the appropriate technical training, followed by attendance at the Air and Space Basic Course. After several years of practical experience in their initial assignments, they will receive further leadership education through the Squadron Officers School. Select Air Force civilian members may similarly attend these education courses.

Effective commanders and supervisors find the proper balance of training, education, and leadership opportunities to develop the tactical competence of their Airmen. They should work closely with educators and trainers and follow the guiding principles below when developing their tactical-level Airmen.

Tactical Guiding Principles

- ✪ **Build Air Force cultural awareness.** Airmen should understand how they function as part of the air and space power team. The credibility gained through experience is of greater relevance if it includes an understanding of how the Air Force operates and what principles of war and tenets of air and space power underpin Air Force effectiveness. Education programs should provide a consistent and systematic promotion of Air Force culture.

- ✪ **Bond Airmen to core values.** Integrity first, service before self, and excellence in all we do—every Airman must understand and internalize these values. They are the core of Air Force leadership and must be taught to Airmen from the start and reinforced at each step of their career through formal and informal education, mentoring, and by the example of Air Force commanders and supervisors as role models.

- ✪ **Build skill competence.** Success on the battlefield requires Airmen to be competent in their skills. This competence directly contributes to "excellence in all we do." No profession is as unforgiving of failure as the profession of arms, especially in the lethal, dynamic, high tech world of today. This is a core training objective.

- ✪ **Build expeditionary expertise.** Once individuals are trained to competence in their career fields, they need to deploy and practice their skills in the dynamic environment in which they will be expected to fight. Training and field exercises develop necessary skill sets for deployed operations. This will enhance their progression from technical competence to specialty expertise within the expeditionary culture. Training programs should be based upon proven principles and lessons learned. Training and education must emphasize an expeditionary approach to tasks and problem solving so Airmen will acquire skills and tactics across the spectrum of conflict.

- ✪ **Build joint and coalition knowledge.** Contemporary military operations involve Airmen in joint operations and very often coalition operations. As they develop their expertise, the skills development process is enhanced by participation in joint and coalition exercises and contingency deployments. Such exercises or real-world events provide awareness of the integration of air and space power with the capabilities of other Services, components, and allies.

✪ **Build expertise through mentoring.** Mentoring offers a skills development multiplier of tremendous value. Through mentoring, senior Airmen can convey experiences and expertise to the more junior. Mentoring offers a means of increasing skill expertise in addition to reinforcing Air Force culture and values. This mentoring can take many forms, from a guest speaker at a formal school or unit gathering, to the "old head" taking the "new kid" under his or her wing. Mentoring establishes trust and loyalty between Airmen and the Air Force, providing developmental opportunities for the individual while displaying concern on the part of the mentor. Role modeling may be considered mentoring, and senior individuals should be aware that their conduct and bearing have an influence on more junior individuals. Mentoring also develops the "Air Force Team" by providing opportunities to talk freely and learn in a more relaxed atmosphere than a formal school can provide.

EDUCATION AND TRAINING AT THE OPERATIONAL LEVEL

During an operation, decisions have usually to be made at once; there may be no time to review the situation or even to think it through.... If the mind is to emerge unscathed from this relentless struggle with the unforeseen, two qualities are indispensable: First, an intellect that, even in the darkest hour, retains some glimmerings of the inner light [commander's vision] which leads to truth; and second, the courage to follow this faint light wherever it may lead.

—Karl von Clausewitz

Education and training at the operational level broaden understanding of integrating expertise to produce operational effects for Air Force missions and continue to build skills. At this level, education assumes a larger role in an Airman's development. It is intended to enhance professional competence through intermediate developmental education. Operational-level education is focused on furthering expertise across related specialties and increasing leadership responsibilities. Operational-level training continues to build tactical skills and develops professional competence.

Operational education and training consist of:

✪ Developmental education

✪ Professional continuing education programs

✪ Advanced academic degree programs

✪ Education with industry

✪ Fellowships

✪ Specialty schools/advanced training

For example, majors will be expected to perform duties as flight commanders or operations officers to gain skills at a higher level in the squadron and complete intermediate service school or a selected graduate-level degree program to further their educational needs as maturing professionals. Noncommissioned officers attend relevant specialty schools and pursue professional continuing education programs. Civilian personnel at this level fill positions with greater organizational, technical, and supervisory responsibilities. As with their military counterparts, they may be selected to attend an advanced academic degree program or education with industry.

(General Carl) Spaatz possessed a good measure of (a) necessary ingredient of a successful general—the ability to inspire trust in both superiors and subordinates. His chief lieutenant, Jimmy Doolittle, in an oral-history interview with Ronald R. Fogleman, then a major, stated, "I idolize General Spaatz. He is perhaps the only man that I have ever been closely associated with whom I have never known to make a bad decision." This praise, coming from a man of enormous physical and moral courage and high intellect, speaks for itself.
—**Dr Richard Davis, published in Aerospace Power Journal, Winter 1997**

General Carl Spaatz, First CSAF, 1947-1948

Operational Guiding Principles

Commanders, supervisors, educators, and trainers must assist Airmen in growing the skills and competencies they need to perform roles at the operational level. They should follow the guiding principles below.

✪ **Hone and enhance Air Force cultural awareness.** As Airmen transition to the operational level they participate in the direction of Air Force operations and deepen their understanding of Air Force culture and the capabilities of air and space power. Education programs should provide a consistent and systematic continuation of that growth. At this point the Airmen should be transitioning from student to teacher of Air Force culture and should seek opportunities to mentor less experienced Airmen.

✪ **Reinforce core values through exemplification.** As Air Force operational leaders Airmen must now consistently and more visibly exercise core values. The Air Force must do its part by reminding, encouraging, and reinforcing this process so that at the next level (strategic) Air Force leaders are the embodiment of Air Force core values.

✪ **Integrate competency and expertise into air and space operations.** At the operational level Airmen integrate numerous Air Force capabilities to accomplish objectives. To effectively integrate those elements and achieve excellence, Airmen must attain a broader competence founded on the tactical skills developed earlier and expanded to include a full understanding of related capabilities. Operational competence mandates a perspective that integrates air and space forces to produce air and space power. Airmen and their commanders should seek opportunities to reinforce, train, and expand these operational skills in the field during exercises or contingencies. Airmen should attend advanced education such as war planning or applications courses. Education and training that emphasizes flexibility of operations and the dynamics of modern warfare prepare Airmen for success across the spectrum of conflict. An example of this integration occurs in the activities of an air and space operations center (AOC), where an Airman must use the competency and expertise gleaned from education, training, and experience in efforts to support air and space operations for a commander, Air Force forces.

- ✪ **Hone joint and coalition understanding.** At the tactical level Airmen are introduced to the capabilities of other Services and nations, but at the operational level they must put what they have learned into practice in an integrated manner to bring success. The Airman's development in this area continues through formal education and training programs, joint and coalition wargames and exercises, and mentoring. Airmen apply this understanding, for example, when operating on the staff of a combatant command.

- ✪ **Expand expertise through mentoring.** At this level, Airmen are positioned to both give critical mentoring and receive significant benefit from it. Airmen should actively seek the wisdom available from more senior Airmen. They should also pass their own wisdom and experience on to their subordinates and less experienced coworkers. Airmen should realize they are seen as role models by lower grades and should act accordingly.

If officers and men believe a general knows what he is talking about and that what he orders is the right thing to do in the circumstances, they will do it, because most people are relieved to find a superior on whose judgment they can rest. That indeed is the difference between most people and generals.
—**Barbara Tuchman, historian and author**

General T. Michael Moseley talking with troops in Iraq

EDUCATION AND TRAINING AT THE STRATEGIC LEVEL

Education and training at the strategic level assists in developing the skills to form accurate frames of reference, make sound decisions, uncover underlying connections to deal with more general issues, and engage in creative, innovative thinking that recognizes new solutions and new options. At this level, education assumes a predominant role in an Airman's development. Education emphasizes understanding of broad concepts and offers insights into complex issues not commonly available in operational environments. It focuses on the institutional Air Force and joint, interagency, business, and international views. Exercises and wargames provide for strategic level training. Strategic development is commonly presented through:

- ✪ Operational assignments

- ✪ Institutional education

- ✪ Self-development

- ✪ Mentoring

- ✪ Exercises and wargames

At this level, assignment to senior command or staff duties at the directorate/division level will round out the skills of the Airman. Attendance at senior developmental education programs, such as Air War College or the Senior Noncommissioned Officer Academy, improves breadth of professional development.

Strategic Guiding Principles

In educating and training senior Airmen in strategic skills and capabilities, the following principles apply:

- ✪ **Leverage experience to further education.** Senior leader education should recognize and be adapted to the experience and competence of the individual. Individuals who achieve senior rank or position within the Air Force bring with them a career-long wealth of experience that should be considered in devising an education plan. Mission needs, especially those of the senior leader's current or immediate future assignment, should drive the programs' selection. For example, a general officer anticipating a NAF commander's job should consider attending a formal course to learn the detailed responsibilities and intricacies of the COMAFFOR and joint force air and space component commander (JFACC). Senior leaders must continue to model the qualities of lifelong learning and personal growth. The concept of the leader as teacher is most critical at senior levels of leadership. Strategic leaders devote a significant amount of time developing the next generation of leaders.

- ✪ **Leverage the senior leader's time.** Time is precious, especially to senior leaders who often must prioritize the many competing demands on their time, choosing to focus on issues that most contribute to meeting their mission requirements or objectives and delaying, delegating, or dismissing the rest. Education programs that incorporate mentorship and self-directed programs along with convenient and complementary participatory and structured education methods will bring the most success due the flexibility brought to bear. At the same time, mentorship is extremely valuable at the strategic level, especially when the senior leader can access the wisdom of another senior leader in a timely fashion when confronting a difficult issue. Senior leaders should remain receptive to providing such mentoring to others, at the strategic and lower levels as all Airmen can also benefit from this guidance.

> *Not only does an air force know what each branch of aviation is doing in the air when acting with an army or with a navy, but it has to keep in constant touch with everything that is taking place on the ground or the water. The airman, therefore, not only from his position of advantage in the sky looks down on the whole field of battle and surveys the contests between hostile air forces in the clouds, but he knows the particular mission which every distinct part of the force, whether on the ground or in the air, has had assigned to it. He knows a great deal about the general operations and mission of the whole force, so that he can act independently on it in case of necessity. Many times the good judgment of a lieutenant pilot has changed the whole aspect of an air and ground battle.*
> **—Brigadier General Billy Mitchell, Deputy Chief of the Army Air Service and early airpower advocate**

✪ **Focus on senior leader skills.** Strategic-level development should focus on the key areas that senior leaders will use. Education and training should hone their ability to express Air Force views within joint, interagency, and international fora. They should be taught to assess the international environment in which they will operate. This assessment should include the strengths, weaknesses, and cultural considerations of the enemy as well as the US and/or coalition forces, and the steps needed to parlay the current situation into a success. Education and training improve their use of processes to meet mission demands and new challenges, and integrates the variety of air and space systems to produce more effective capabilities. Senior leaders are taught how to make informed decisions and how to drive transformation and execution to meet the demands of dynamic environments. A successful program will enable them to align their organization to serve the personnel, the Air Force, and the nation, and shape the way air and space forces are employed. Education and training programs should incorporate Air Force culture and core values.

CONCLUSION

If we should have to fight, we should be prepared to do so from the neck up instead of from the neck down.

—General James H. Doolittle

General Doolittle leading the 18 April 1942 raid on Tokyo.

Leadership is fundamental to the United States Air Force. Creating future Air Force leaders is the responsibility of the current leaders, and force development is their tool to do so. The more effort that is placed in using this tool, the better the leaders it will produce. By using the organized approach of developing leaders from the tactical level, through the operational, leading to the most senior strategic levels in the Air Force, the Service will ensure its continued preeminent position in the world. Leaders are inextricably linked to mission effectiveness; developing those leaders in a deliberate process that guarantees the Air Force will produce the requisite leadership. **Leadership and force development must continue to provide the Air Force with its most valuable resource: its people, its motivated and superbly qualified Airmen.**

At the very heart of warfare lies doctrine....

SUGGESTED READINGS

AIR FORCE DOCTRINE DOCUMENTS

All Air Force personnel should be familiar with the full breadth of Air Force operations. As a beginning, they should read the entire series of the basic, capstone, and keystone operational doctrine documents.

- ✪ AFDD 1, *Air Force Basic Doctrine*

- ✪ AFDD 2, *Organization and Employment of Aerospace Power*

- ✪ AFDD 2-1, *Air Warfare*

- ✪ AFDD 2-2, *Space Operations*

- ✪ AFDD 2-3, *Military Operations Other Than War*

- ✪ AFDD 2-4, *Combat Support*

- ✪ AFDD 2-5, *Information Operations*

- ✪ AFDD 2-6, *Air Mobility*

- ✪ AFDD 2-7, *Special Operations*

- ✪ AFDD 2-8, *Command and Control*

JOINT PUBLICATIONS

- ✪ JP 0-2, *Unified Action Armed Forces (UNAAF)*

- ✪ JP 1, Joint Warfare of the Armed Forces of the United States

CHIEF OF STAFF OF THE AIR FORCE READING LIST (2003)

The Chief of Staff of the Air Force Reading List is available for review at: http://www.af.mil/lib/csafbook/readinglist.shtml.

CATEGORY I: History of the Air Force from its beginning through its major transformations as an institution.

Boyne, Walter J., *Beyond the Wild Blue: A History of the United States Air Force 1947-1997* (St. Martin's Press). 1997.

Copp, DeWitt S., *Frank M. Andrews, Marshall's Airman*, (Air Force History and Museums Program). 2003.

Lambeth, Benjamin S., *The Transformation of American Air Power*, (Cornell University Press). 2000.

Perret, Geoffrey, *Winged Victory: The Army Air Forces in World War II*, (Random House). 1993.

CATEGORY II: Insight into ongoing conflicts and the frictions that can produce conflicts in the future.

Huntington, Samuel P., *The Clash of Civilizations and the Remaking of World Order*, (Simon and Schuster). 1997.

Lewis, Bernard, *The Crisis of Islam: Holy War and Unholy Terror*, (The Modern Library). 2003.

Margolis, Eric S., *War at the Top of the World: The Struggle for Afghanistan, Kashmir, and Tibet*, (Routledge). 2001.

Meyer, Karl E., and Shareen Blair Brysac, *Tournament of Shadows, The Great Game and the Race for Empire in Central Asia*, (Counterpoint). 1999.

Yergin, Daniel, *The Prize: The Epic Quest for Oil, Money, and Power*, (Simon and Schuster). 1990.

CATEGORY III: Organization, leadership, and success stories holding lessons for the present and future.

Creech, Wilbur L., *The Five Pillars of TQM: How to Make Total Quality Management Work for You*, (Truman Talley Books/Dutton). 1994.

Puryear, Edgar F., *American Generalship Character is Everything: The Art of Command*, (Presidio Press). 2000.

CATEGORY IV: Lessons emerging from recent conflicts and the preparation for them.

Clancy, Tom, with General Chuck Horner (USAF Ret.), *Every Man a Tiger*, (G.B. Putnam's Sons). 1999.

Cohen, Eliot A., *Supreme Command: Soldiers, Statesmen, and Leadership in Wartime*, (The Free Press). 2002.

Kitfield, James, *Prodigal Soldiers*, (Simon and Schuster). 1995.

APPENDIX A

USAF OATH OF OFFICE/ENLISTMENT

USAF Oath of Office/Enlistment

Why we administer an Oath of Office and Enlistment:

1. **Legal Requirements.** Federal law requires persons enlisting in the Armed Forces, or persons elected or appointed to a position of honor or profit in the government of the United States, to subscribe to an oath before beginning in the position.

2. **Taking the Oath.** People being appointed or commissioned in the Regular Air Force, Reserve of the Air Force, or US Air Force (Temporary), or being appointed as a civilian employee, must execute the oath of office/enlistment when they accept the appointment or commission. Very simply, an oath is a promise—an ethical agreement or bond of a person's word.

3. It is with these oaths that we, as Airmen, first commit ourselves to our basic core values, placing service to our Constitution, our President, and our compatriots before ourselves. It is where we place integrity on the line by giving our word as our bond. And it is where we swear (or affirm) to "well and faithfully" discharge our duties, or obey orders to do so, thus committing ourselves to excellence.

4. An individual, except the President, enlisting in, or elected or appointed to an office of honor or profit in the civil service or uniformed services, shall take the appropriate oath:

OFFICER AND CIVILIAN	ENLISTED
I (FULL NAME),	I (FULL NAME),
HAVING BEEN APPOINTED A (Grade in which appointed) IN THE UNITED STATES AIR FORCE,	
DO SOLEMNLY SWEAR (OR AFFIRM) THAT I WILL SUPPORT AND DEFEND THE CONSTITUTION OF THE UNITED STATES AGAINST ALL ENEMIES, FOREIGN AND DOMESTIC; THAT I WILL BEAR TRUE FAITH AND ALLEGIANCE TO THE SAME;	DO SOLEMNLY SWEAR (OR AFFIRM) THAT I WILL SUPPORT AND DEFEND THE CONSTITUTION OF THE UNITED STATES AGAINST ALL ENEMIES, FOREIGN AND DOMESTIC; THAT I WILL BEAR TRUE FAITH AND ALLEGIANCE TO THE SAME;

AND THAT I TAKE THIS OBLIGATION FREELY, WITHOUT ANY MENTAL RESERVATION OR PURPOSE OF EVASION;	
AND THAT I WILL WELL AND FAITHFULLY DISCHARGE THE DUTIES OF THE OFFICE UPON WHICH I AM ABOUT TO ENTER,	AND THAT I WILL OBEY THE ORDERS OF THE PRESIDENT OF THE UNITED STATES AND THE ORDERS OF THE OFFICERS APPOINTED OVER ME, ACCORDING TO REGULATIONS AND THE UNIFORM CODE OF MILITARY JUSTICE,
SO HELP ME GOD.	SO HELP ME GOD.

APPENDIX B

ENDURING LEADERSHIP COMPETENCIES

Personal Leadership

✪ **Exercise Sound Judgment:**

✪✪ Develop and apply broad knowledge and expertise in a disciplined manner when making decisions

✪✪ Taking all critical information into account, considering interrelationships between issues and the implications for other Air Force stakeholders

✪ **Adapt and Perform Under Pressure:**

✪✪ Personally manage change and maintain continuity for self and others when mission requirements, work tasks, or processes change

✪✪ Maintain composure and continue to work constructively and resourcefully under pressure

✪ **Inspire Trust:**

✪✪ Maintain high standards of integrity

✪✪ Establish open, candid, and trusting relationships, and treat all individuals fairly and with respect

✪✪ Subordinate personal gain to the mission's success and demonstrate loyalty to the unit and the chain of command

✪ **Lead Courageously:**

✪✪ Display both moral and physical courage by showing a willingness to take risks, act independently, and take personal responsibility for actions

✪✪ Persist with focus and intensity even when faced with adversity

✪✪ When challenged, project confidence, credibility, and poise

✪ **Assess Self:**

✪✪ Understand how personal leadership style and skill impact decisions and relationships with others

✪✪ Create a personal leadership development plan using insight gained from assessing values, personal strengths and weaknesses along with performance preferences and learning style

✪✪ Apply insight and learning to improve leadership performance

✪ **Foster Effective Communication:**

✪✪ Ensure a free flow of information and communication up, down, across, and within an organization by actively listening and encouraging the open expression of ideas and opinions

✪✪ Express ideas clearly, concisely, and with impact

Leading People/Teams

✪ **Drive Performance through Shared Vision, Values, and Accountability:**

✪✪ Instill commitment to a common vision and shared values

✪✪ Create a climate that fosters personal and organizational excellence

✪✪ Set high expectations for performance and convey confidence in others' ability to achieve challenging goals and overcome obstacles

✪ **Influence through Win/Win Solutions:**

✪✪ Promote ideas, proposals, and positions persuasively through compelling rationale and arguments

✪✪ Consider underlying consequences for key stakeholders while seeking and negotiating win/win solutions

✪ **Mentor and Coach for Growth and Success:**

✪✪ Guide the development of junior members toward development of talents, skills, and aspirations to achieve a successful Air Force career

✪✪ Appraise individual strengths and weaknesses, provide constructive feedback, reinforce efforts and progress, identify career opportunities, and champion individual success

✪ **Promote Collaboration and Teamwork:**

✪✪ Facilitate and encourage cooperation among team members

✪✪ Recognize and share credit for success

✪✪ Work as needed with peers and subordinates to establish a group identity through mutual goals, common team practices, and structure

✪ **Partner to Maximize Results:**

✪✪ Proactively cultivate an active network of relationships inside and outside the Air Force

✪✪ Accommodate a variety of interpersonal styles and perspectives to achieve objectives and remove barriers

✪✪ Leverage cross-disciplinary knowledge to provide integrated solutions

Leading the Institution

✪ **Shape Air Force Strategy and Direction:**

✪✪ Establish critical long-range success factors and goals designed to achieve mission and organizational advantage

✪✪ Use an understanding of key economic, social, and political trends both domestically and globally to assess strengths, weaknesses, opportunities and threats to develop strategy

✪✪ Energize the organization through a compelling picture of the future opportunities the Air Force has to offer

✪ **Command Organizational and Mission Success through Enterprise Integration and Resource Stewardship:**

✪✪ Effectively prioritize, manage, and integrate diverse mission elements across varying environments to address the situational requirements associated with the required response, access to resources, and deployment of people, equipment, supplies, technology, and funding essential to organizational and mission success

✪✪ Develop, implement, and refine a thorough operational risk management strategy to enhance warfighting capability by effectively preserving personnel and resources

✪✪ NOTE: Command is a legal authority, not exclusively a moral or ethical one

✪ **Embrace Change and Transformation:**

✪✪ Lead efforts to streamline processes and adopt best practices

✪✪ Create an environment that supports innovation, continuous improvement, and risk taking

✪ **Drive Execution:**

✪✪ Translate strategies into operational results by identifying supporting goals and tasks along with individual accountabilities

✪✪ Align communication, people, processes, and resources to drive success

✪✪ Ensure that measurement systems are in place to track effective implementation and results

✪ **Attract, Retain, and Develop Talent:**

✪✪ Assess capability and talent needed to propel organizational and individual performance

✪✪ Build leadership bench-strength by ensuring that systems are in place to attract a high caliber, diverse work force

✪✪ Retain top talent over time by creating an environment that encourages personal achievement, continuous learning, creativity, and promotional opportunities

✪✪ Address career and work-environment issues that affect retention, including physical and mental health

APPENDIX C

EDUCATION AND TRAINING

The event that facilitates the transition from one level of experience to the next is developmental education in preparation for a developmental assignment. Education and training are critical components of the force development construct in preparing individuals to gain productive experiences. **The success of the force development construct depends on relevant and appropriate education and training.** Although both education and training are essential to operational capability, they are fundamentally different. Education prepares individuals for dynamic operational environments, while training is essential in developing skill sets for complex systems. Education and training are complementary and will commonly overlap, and while the distinction between them is unimportant within this 'gray area,' the distinction between their essential natures remains critical to the success of each. The following items distinguish education from training:

✪ **Training** is appropriate when standardized outcomes are required. Training is focused on building specific skill sets to produce reliable, consistent results. Although skill application involves judgment, it is the purpose of training to teach skills that are associated with desired outcomes. (When repairing jet engines, for example, it is desirable to have the engine meet standardized performance measures upon completion of the repair tasks—proper training ensures a standardized, predictable outcome.)

Education is appropriate when adaptive outcomes are desired. Education is focused on developing critical thought that enables creative solutions. Although creative thought may involve skill application, it is the purpose of critical thought to form successful solutions to new problems. (An engineer, for example, is able to design a new jet engine that exceeds all known performance measures through application of creative design concepts and unusual materials applications.)

✪ **Training** is task dependent. Training is generally focused on a specific skill. Although specialties may be quite complex, each is composed of elements having distinct tasks that when correctly performed lead to successful, predictable outcomes. (When operating a radar system, for example, power-up is distinct from data recall operations and each task is required, but each task is distinct in the steps taken—proper training ensures the proper steps are followed in the proper sequence to successfully operate the system.)

Education is process dependent. Education is generally focused on combining familiar and unfamiliar information to produce a suggested course of action. The intellectual demands of consolidating past experiences and ideas with new experiences and unfamiliar information to produce new ideas depend on the process of critical thought. (Radar data, for example, only becomes useful when analyzed and screened to produce relevant information—the intellect required to make sense of multiple items forms the process of critical thought.)

✪ **Training** is technically specific. Training is intended to develop skill sets that are associated with particular duties. While some skill sets are generally universal (such as computer skills), specialty training is specific to a particular skill set. Skill sets are generally associated with specific duty requirements and the tools of that specialty. Training is focused on specific situations and the tools of that specialty. (A tanker boom operator, for example, would not be prepared to take on the duty requirements of a para rescue Airman or vice versa -- each has received training that is specific to the technical tools of his or her duties.)

Education is not dependent on situation. Because education seeks to develop critical thinking skills, it attempts to prepare individuals for new experiences and new challenges. While education can readily prepare individuals for known situations, the fundamental aim is to develop individual talents to create successful outcomes in unfamiliar situations. The goals of education do not depend on a specific situation to produce success. (Using a comprehensive understanding of atmospherics, a weatherman for example, is able to accurately predict weather patterns across global regions.)

✪ **Training** requires restrictive application. Since training is generally focused on a specific skill set, the skills learned are usually limited to the specialty related to that skill set. Training aims to instill certain specific skills that when applied in a systematic and predictable way produce predictable outcomes. Training is, therefore, generally restricted in application to the known circumstances related to the skill set. (A jet engine mechanic, for example, would not be well equipped to repair a piece of complex communications equipment any more than a electronics technician could be expected to repair a jet engine.)

Education requires transformative application. Since a goal of education is to instill critical thinking skills, it is in unfamiliar circumstances that education can have the greatest benefit. Education provides the individual with logic skills that encourage creative thought and allows individuals to create new solutions to unfamiliar problems. In application, education is most beneficial when transitioning from the known to the unknown, thus is best suited to transformative application. (An engineering team, for example, confronted with the task of repairing an unfamiliar foreign communications network devises a successful solution using equipment that is both familiar and foreign.)

✪ **Training** functions best within defined parameters. Training develops skill sets and the talent to successfully cope with deviations from normal, within the bounds of the specialty.   Training is skill specific and variations from those normally expected circumstances are also limited to that skill set. (An aircraft hydraulics specialist, for example, is trained to deal with hydraulic systems and expected problems, but would likely not be as successful in coping with an unfamiliar hydraulic system that experiences an unfamiliar failure.)

Education functions best outside defined parameters. The essential strength of education is to prepare individuals to create successful outcomes in unfamiliar situations. The value of education is most apparent when the individual is confronted with creating solutions beyond the set of parameters in which they may normally operate. (A hydraulic specialist, for example, relying on an understanding of hydraulic principles and system functions is able to create a solution to an unfamiliar failure.)

✪ **Training** functions best within expected environments. Training generally serves to impart skills within known and likely operating environments. Circumstances that are known as normal operating environments and situations that can be anticipated may be considered as occurring within an expected environment. Training provides the skills necessary for success in stable environments. (Emergency drills and realistic exercises, for example, help develop skills to cope with anticipated scenarios under stress or in critical situations.)

Education functions best within unexpected environments. In unexpected or unanticipated situations there are no procedures or checklists to provide guidance. Skill sets generally become less applicable in scenarios that have not been seen or practiced. Education provides the tools necessary to cope with new challenges. It is in rapidly changing environments that produce unexpected problems that education can provide the mental talents to succeed. (A fire fighter, for example, when being confronted with unmanageable flames understands the mechanics of fires to successfully egress the situation.)

✪ **Training** value diminishes with uncertainty. The further the situation progresses from the talents of the individual, the less effective the individual becomes in implementing a successful solution. Because training is focused on a specialized skill set, those circumstances that fall outside of the skill set produce a greater amount of uncertainty. Thus the value of skill set training is reduced in the face of uncertainty. This could be likened to a 'fish out of water.'

Education value increases with uncertainty. Education provides the tools for innovation and creative thought. In circumstances of new challenges and unfamiliar situations, education can allow individuals to create solutions to reduce uncertainty and implement successful solutions. (Combat presents leaders with many opportunities to experience unfamiliar situations, but relying on historical precedents, lessons learned in wargames and exercises, and past personal experience leaders can develop successful strategies and tactics to prevail.)

✪ **Training** is not inherent in education. Learning can take place in individuals having few specialized skills. Even in unstructured environments, learning can proceed successfully. Education involves the process of teaching individuals new concepts and/or developing logic talents to create new thought. There are many examples of successful artists creating great works without a formal training in the medium. It is creative talent that is among the most beneficial results of education.

Education is inherent in training. Basic talents are critical to learning. Individuals, for example, must be able to read proficiently to access training materials. Individuals must also have a good grasp of vocabulary to understand training terms and concepts. Subjects such as reading, vocabulary, mathematics, and similar topics are the product of education. Training cannot take place without first having individuals who meet the qualifications to receive training. Training that exceeds the qualifications required of the participants is less effective.

✪ **Training** shows immediate benefits. Learned skills can be demonstrated almost immediately. It is often part of the training process that individuals demonstrate the skills acquired. Repetition of skills serves to reinforce those skills and provides a measure of training success. A 'three-level' technician can be placed in positions of responsibility and produce successful outcomes as a result of training. Training usually produces immediate effects by imparting new skills or developing existing skills.

Education provides long-term benefits. Skills in critical thinking are usually not demonstrated until encountering unfamiliar circumstances. Logic skills are also developed over time through formal education and experience, thus are constantly evolving and maturing. Consequently, the benefits of education tend to grow over time closely linked to experience and appear more often as a benefit some time after the education process. Because logic skills are not as demonstrable as technical skills, these talents are usually not as apparent in the short-term.

APPENDIX D

CASE STUDIES

The following case studies illustrate the breadth of leadership the Air Force has been privileged to experience over the course of its existence. They portray leaders using the skills they learned as officers and enlisted personnel to enhance the development of the Air Force. The first study, the First Chief Master Sergeants of the Air Force, details, in their own words, how that office was created by the men who lived its earliest formative era. The second study, General Creech and the Transformation of Tactical Air Command (TAC), examines a senior officer's efforts to create the most efficient major command he could for the Air Force. The third study, General Ryan and the Creation of the AEF, details the efforts by the Chief of Staff of the Air Force to produce an expeditionary mindset throughout the entire Service.

The case studies are intended as examples of leadership and force development. They are presented in chronological order. Many additional studies can and, no doubt, will be done. Military educational organizations and students should take advantage of the opportunity to expand their knowledge on leadership and force development by researching and writing additional case studies.

Case Study: The First Chief Master Sergeants of the Air Force

Creation of the office of the Chief Master Sergeant of the Air Force was a dramatic step forward in leadership and force development for the Air Force. The first individuals to hold that position provide unique insight into that process. Their own words most effectively communicate the challenges and successes the creation of that position offered the Air Force. It demonstrates the necessity of developing a force with an active officer/enlisted relationship.

The following interview was conducted 24 June 1987 by Dr. Richard H. Kohn, Chief of the Office of Air Force History. It is excerpted from *The Enlisted Experience: A Conversation with the Chief Master Sergeants of the Air Force* (Air Force History and Museums Program, 1995).

Dr. Kohn interviewed four of the first five Chief Master Sergeants of the Air Force (CMSAF) for this work: Paul W. Airey, Donald L. Harlow, Thomas N. Barnes, and Robert D. Gaylor. (Note: CMSAF Richard D. Kisling, 1 October 1971-30 September 1973, the third CMSAF, passed away on 3 November 1985.)

ESTABLISHING THE CHIEF MASTER SERGEANT OF THE AIR FORCE

Kohn: You've brought up the Chief Master Sergeant of the Air force position. It was established in 1967, at the beginning of these years of turmoil. Why was it created? How was it created? What was the need for it? Why was there opposition to it?

Paul W. Airey (First CMSAF, 3 April 1967-31 July 1969): A lot has been said about this position. Of course, the basic job description of the Chief Master Sergeant of the Air Force is to aid and advise the Chief of Staff and the Secretary of the Air Force in all matters pertaining to enlisted personnel. That job description is still in effect. Actually, I think you hit it, Dr. Kohn, by linking it to the turmoil of the 1960s. We were at the start of a period when the leadership realized suddenly that they needed better communications. Many people say that's the reason this job was established: to give enlisted people a route right to the top without going through the various channels, so that they would have somebody up there representing them. And I think that's correct. But I also think that one of the unwritten aspects—one I feel so very strongly about—is that it's a position that all enlisted people can look up to and say, "Maybe someday I could be the Chief Master Sergeant of the Air Force." To me, that means just as much, or counts for just as much, as anything else.

Donald L. Harlow (CMSAF, 1 August 1969-30 September 1971): I've always said that when Paul (Airey) was in the position, Air Force leaders didn't really want the office…. [They] didn't want it because [they] felt that we were trying to set up a separate communication chain.
Kohn: One that would undermine the formal chain of command?

Harlow: Yes. One that would undermine the formal chain of command. So there was an awful lot of opposition. Even CMSAF Dick Kisling ran into it. It wasn't until later on that the leadership started to realize that it was a good position. We had no authority; we couldn't sign anything. When Paul and I were asked to comment on various issues, usually through staff summary sheets, we gave our input. As each one of us got into the position, the job increased in importance and significance. Those who followed served on more committees and got to go over to Congress to testify on various issues. The position became more visible, and I think that was great.

I think that the Senior Enlisted Advisor program, when it started off, was not what it is today for the simple reason that the commander, whether he wanted it or not, thought, "Well, I'll have one." Some of the senior advisors went out to do the job, and for some it was great to be a Senior Enlisted Advisor, going to all the banquets, the social functions, and things like that. But they weren't really into the program of doing something for the troops, or listening to the troops. With time, this position has increased in significance. There's an indoctrination course now for Senior Enlisted Advisors. I think the position itself is a very good one and very important today.

Thomas N. Barnes (CMSAF, 1 October 1971-30 September 1973): I'd second those comments. I'd like to go back to the initial perception and subsequent acceptance of the position in terms of how each Chief of Staff developed it. Specifically, when General George Brown discussed it with me, he said, "There are places where it will be cosmetically good for you and me to appear together. I'll identify those." That cut through the ice for me. He'd identified the pockets of resistance that apparently had surfaced. I saw where there would've been a great problem of my gaining acceptance had he not done that. Then, when he went on to become the Chairman of the Joint Chiefs of Staff and General David Jones came in as the Air Force Chief of Staff…he had a broader perspective in that he had a way of explaining it to people. So, for me the ice had been broken.

Paul and Don and Dick had made the initial strong steps, so that, fortunately, you didn't have to dodge anything or put your head down when you walked in. I think, as the position developed and grew, each of us was able to increase that respect a little bit. We found, however, that there were a whole series of things to do, and that there are some things that have run continually through each of our tenures and they still aren't fixed. These things are perhaps beyond the purview of this position to fix, and they aren't going to be fixed until other matters are resolved.

By and large, the message to the enlisted force was that there was a representative at the Air Staff level who had some exposure in the policy-making shop. I think there exists a perception in the enlisted force that, "If I call the Chief and talk with him, things ought to change overnight." That's hardly going to happen, but there are things that I think can be pointed to that have changed directly as a result of actions by the men in this position. Each of us can name things that occurred during our tenures which changed fairly quickly as a result of our input to the Air Staff.

I think we maintained a cleaner slate, perhaps than the other Services in this regard. I think the position has been better used by the Air Force Chiefs of Staff, than perhaps the Chiefs of Staff of the Army. I say that because of how the position has been used—as a staff level position as opposed to a position of authority. It's not authoritative. It doesn't establish authority, but it participates in policy formulation at certain points.

Kohn: Chief Airey, when you started as the first Chief Master Sergeant of the Air Force, did you have an agenda? When you took the job in 1967, were you given an agenda?

Airey: Not really. I think the Chief of Staff was watching me and waiting for reports on me. It took about six months, and from then on, I couldn't ask for a better supporter. It's generally well known that General John Paul McConnell didn't really care about the position's being established. I was told by him to make my own agenda. I was told, "Don't upset the Air Staff." But the agenda was left to me, and I was given a pretty free hand.

Robert D. Gaylor (CMSAF, 1 August 1977-31 July 1979): I'm glad it happened that way. I think all of us had to establish our own agenda, I really do. Had it been dictated to us, we would've been merely pawns or mouthpieces of the Air Staff.

Harlow: And we would have been frustrated.

Gaylor: Yes, eventually we would have, so I'm glad they did it that way. I'm glad they said, "Okay, Chief Airey, go for it and we'll watch you."

Airey: I think I was observed and reports were sent back.

Gaylor: We all were. Constantly. We were in a fish bowl.

Kohn: In its formative stages was the Chief Master Sergeant of the Air Force caught in an ambivalent position between representing management—specifically, the Chief of Staff and the "system"—on the one hand and on the other representing the enlisted force?

Airey: I made a statement once, and it's been quoted—even misquoted—several times. I said that I wasn't going to be a clearinghouse just for enlisted gripes. I was there to help the enlisted force with their problems and to try and help rectify those problems if they were bona fide, but I still expected people with the everyday complaints to go to their First Sergeant and their supervisor.

Kohn: You wouldn't be an ombudsman?

Airey: No.

Harlow: A lot of people ask me now in my travels, "What did you recommend when you were in the job?" You can't reveal to the field what you recommend, for the simple reason that if it worked, fine, but if it didn't work, they'd say, "You're not effective. You made this recommendation, and nothing has happened."

Barnes: I think one of the most, if not *the* most, challenging aspect of the job, was maintaining credibility with the enlisted force and credibility with the Air Staff. There was a middle ground you were placed in which required both. Unfortunately, this point was invisible to the enlisted force. [They didn't see] you in the same light that the Air Staff saw you. But I think we all had the duty to touch those things that directly impacted the day-to-day lives of the enlisted force. Specifically, we were members of the Uniform Board. Very few things happened in uniform changes that we didn't have a profound impact on. We were also members of the Army Air Force Exchange Board. We were able to voice our opinions; and we saw the utilization of the money that went back into the Morale-Welfare-Recreation system as well. We had a profound impact on a lot of invisible things. Those activities often just went unnoticed.

There were the opportunities to initiate things. There were review activities associated with the budget cycle where the hardware and the people issues occurred. The Chief Master Sergeant of the Air Force was not excluded. You went in and had a chance, if nothing else, to say your piece and understand the process. Also, you went to some of the classified briefings at the morning staff meetings that the Chief of Staff went to and saw the world in his mirror. Then you were debriefed on all those things that you couldn't talk about outside that meeting or, in fact, after your retirement. In short, there was a totality of activities and programs that you understood. You went out and sold things, like Paul said, that you didn't necessarily support in your own mind, but that you did support in the larger picture.

That was the most strenuous part of the job. We all have felt the strain associated with what actually happens to a guy when he knows all of the activities and risks associated with managing the Air Force. It would be traumatic for us to go out and discuss some of the things that we knew for a fact. I think somewhere in the process of selecting people for this job, somewhere in somebody's wisdom, a perspective or concern existed. The Chief of Staff has to select people who can handle large issues, and who don't break under pressure.

Gaylor: One thing I had to learn, and now that I look back, it was interesting. I worked for the Chief of Staff, so you'd think that on returning from a trip you'd brief your boss. So, down the hall I go with this knowledge that I've accrued visiting four bases and share it with the Chief of

Staff. That makes sense. What you don't know is that the Chief of Staff then calls the Director of Personnel on the phone and says, "Get down here. What do you plan to do about this problem?" Later in the afternoon my phone rings, and the Director of Personnel says, "Hey, Chief. Why didn't you let me know? Whenever you go into the Chief of Staff, my buzzer rings. Give me a little advance notice." I say, "That makes sense. I want to work with you, not against you."

So, now I return from another trip armed with this knowledge, and the question is, "Who should I share it with first?" The IG [Inspector General] wants to know it; the LG [Director of Logistics] wants to know it; the DP [Director of Personnel] wants to know it; the Vice Chief of Staff wants to know it; the Chief wants to know it, too. Who do I give it to first? That's a lesson that has to be learned.

Then you've got your staff officers. CMSgt J. B. Wood came to me and said, "Bob, can I level with you? You're causing a problem. We're getting this stuff in the back door. All you've got to do is pick up the phone and alert us to it, and that way when the Chief yells, we're already working the issue. We look good, and you look good." So I learned. I came back from another trip and I've got this head full of knowledge, and I'm asking, who should I tell it to first. You want to get on the loudspeaker and say, "Attention! Attention everyone!" But you have to learn that even though you work for that Chief, you'd better be careful that what you share with him won't in some way undermine the Air Staff or you'll lose their support. They'll hang you out to dry, baby, and you're dead on the vine! I learned that in about a month.

Barnes: You've got to work with that staff. That is an absolute must.

Kohn: Does the Air Staff listen?

Barnes: Yes, they do.

Airey: Yes.

Harlow: Yes.

Gaylor: Because they know you, and could go to the Chief if they don't! You've got a hammer. They appreciate that you come to them first. I used to say to General Bennie Davis, "I've not talked to anyone about this; I wanted to share it with you first." I always got action. They appreciate your doing that so they don't get "backdoored."

Kohn: Can I ask you all what was your greatest satisfaction and your greatest frustration in the job of Chief Master Sergeant of the Air Force?

Gaylor: My greatest achievement was knowing I did the very best I could. I left the job with my head high saying, "You got the best Bob Gaylor had to offer. You may not agree it was the best, but it was the best. You got my total commitment, my total energy. I hit the ground running and never stopped. What more can you ask of anyone?"

My greatest frustration was wishing that I could've done more. Time and distance were my enemies. I wanted to go everywhere, be omnipresent, ubiquitous, and peripatetic. I was unable to be those things because of the time limitation. You look back and say, "I did the best I could; now let somebody else do it."

Harlow: During my watch, anytime I visited a command I usually had a chief master sergeant as an escort. I always sat down with him before I left the command and wrote a report. A copy of that written report went to the major air commander and the Chief, with copies to the Air Staff, on any problems related to that visit. I knew in my mind, even though nobody told me, that when I left that command, anything I said or anything I liked or disliked about that command, someone was going to call the Chief. I wanted to make sure that the commander over there in the Pacific or Europe understood that what I put in the report was going to the Chief and there was no question about that. I felt that was something I had to do to be fair.

Airey: I think that's probably one of the most asked questions: "What do you feel you accomplished?" I feel a lot like Bob Gaylor; you give it your best shot. I'd say if I had to look back and pick out one single accomplishment, I'm proud of the fact that I was on the ground floor of the Weighted Airman's Promotion System. If you ask me, I'll put it this way: Of all the things I was involved in, recommended, suggested, and fought for, I've lost more battles than I've won.

Barnes: For me, obviously, the high point was my two extensions, and principally the issues around which the extensions were necessary. I had been involved in those issues mentioned earlier as they had unfolded and I continued to work them. That was very satisfying. That period from 1975 to 1977 was very frustrating. [We tried] to address what was a downhill slide in enlisted programs and certainly in morale. That was the toughest sledding because the Airmen asked, "Why?" The Chief had been valiant in his efforts to do all he could with morale, and you defended the policies every way you could, but there just was no way to address it.

Airey: I've used this line many times, "Ask not what the United States Air Force can do for me, but what I can do for it. What's my duty and responsibility to the United States Air Force, to my country, and to the enlisted force?"

Kohn: I have a general question about the change during your careers of the relationship between airmen and NCOs, NCOs and officers, and how you established credibility with young airmen.

Harlow: I firmly believe, and I've always believed, and I think the others share my feelings, that "familiarity breeds contempt." I don't believe that you must get so close to an officer that you start calling him by his first name, even if you play golf with him. All enlisted people have to appreciate that and adhere to it. That's part of the discipline.

In the corporate world of today you have some of your junior executives and some of your senior executives who bring all employees in for a little reception or a little party. They can do that, but they don't violate that Mr. So-and-so or Miss So-and-so protocol. They still don't do that, and I think this is, again, a reflection on the military.

Barnes: If we're going to have a military Service, it has to be that way. There's a limit to what relationships can be if you hope to nurture and continue that respect. [The Service] is built on admiration and respect, which is earned. Earning it is the key for those who would command, but I don't think we can ever get away from it and have a successful military force.

Airey: I concur fully. Let me make this historical analogy. In World War II a handful of regulars led us, but the war was really fought by amateurs. We expanded from 20,000 people in 1939 and 1940 to 2.5 million in finally winning the war. Of course, the people who fought it were draftees and enlistees from all walks of life. In my entire bomb group I think there were two regular officers and a handful of regular NCOs. The point I'm getting at is there was a different type of leadership needed then.

Today, the United States Air Force is professional. An airman comes out of basic training and then tech school, and we start sending him to leadership school and to the NCO academies. So the entire force, for the most part, is professional as opposed to amateur. These people need different leadership. We don't need that harsh: "Do as I tell you, or we're going to kill you." I've actually seen people get slapped around and beaten up in my time. But we still need discipline; I'm a great believer in non-fraternization, and when I say "non-fraternization," I keep officers at arm's length, even to this day. I don't call any officer by his first name. If he's a lieutenant, I call him lieutenant. I think in many ways the NCO force is far more intelligent and the Air Force more people-conscious than the Air Force I knew many, many years ago.

Gaylor: We surely didn't "back into" that professionalism. It was developed. I think we are simply reaping the benefits. We now cannot rest on our laurels, but must continue to improve our professionalism. That, to me, is the name of the game.

Harlow: We're grooming the younger people much better in many respects.

Airey: If I had to take a hundred chief master sergeants at random today versus a hundred chief master sergeants back in 1959 and compare them, there would be no comparison when it comes to overall education, overall attitude toward helping people, and in many other aspects. The hundred chiefs today would surface as better leaders overall, but this in no way detracts from the many great NCOs of the past.

Kohn: So we're a stronger force?

Airey: Yes. With all our problems we're a stronger force today.

The above interviews demonstrate the vital role Airmen play in development of the force. These first Chief Master Sergeants of the Air Force created a position with far-ranging effects on the entire Service. In so doing, they enhanced the officer/enlisted relationship, making the Air Force a more cohesive organization. They provided a leadership vector that has allowed continuing advancement of the Air Force's capabilities and did so in a manner to positively impact the officer corps, the enlisted corps, and the Department of the Air Force civilians.

Case Study: General Creech and the Transformation of Tactical Air Command (TAC)

General Wilbur L. "Bill" Creech was the commander of TAC from 1978 to 1984. He stood on the shoulders of previous commanders to create a stronger and more vibrant command. Under his tutelage, Red Flag exercises, developed in answer to training problems identified in Vietnam, spawned a number of related offshoots. A biennial Maple Flag program at Cold Lake, Canada, brought together selected Air Force, Navy, Marine, and Canadian fighter units for a week of intensive ground attack and air-to-air training. Moreover, Canadian, British, German, and other allied aircrews were regularly invited to participate in greatly-expanded Red Flag exercises at Nellis Air Force Base (AFB), Nevada, out of a recognition that US aircrews would fight in a coalition context in any future war. Blue Flag, a nonflying activity conducted at Eglin AFB, Florida, concentrated on the myriad details of large-force mission employment planning. Checkered Flag became an exercise in which each fighter unit throughout the tactical air forces (TAF) planned and exercised for its real-world contingency tasking. Finally, starting in 1985, Copper Flag at Tyndall AFB, Florida, offered a thrice-yearly air defense exercise for aircrews and weapons controllers featuring realistic counterair scenarios not available at their home stations.

Return to Higher Altitudes

Recognizing the multiple demands associated with surviving and performing effectively in the low-altitude arena, TAC's leadership starting in 1978 under General Creech stepped out boldly to seek an appropriate blend of technology and tactics that might help pilots return to higher altitudes where they could escape the dangers of the low-altitude regime and improve their chances of successful target attack. During earlier Red Flags, all starting scenarios each day presumed that it was the first minute of the first hour of a war against undegraded Warsaw Pact air defenses. No kill removal was provided to account for surface-to-air missiles (SAMs) destroyed in previous missions, and low-level penetration to target was invariably the standard practice, on the premise that radar-guided SAM's could not be negated from higher altitudes. Not only did the resulting simulated loss rate to enemy antiaircraft artillery (AAA) and short-range infrared SAMs soar to a point where many pilots concluded that they could not survive in actual combat, the actual aircraft accident rate rose dramatically as a consequence of the unforgiving nature of the training environment. During the first two years of Red Flag, more than 30 heavily task-saturated aircrews lost their lives as a result of either having flown into the ground inadvertently while maneuvering to avoid getting locked up by a simulated threat radar or having collided in midair during a maneuvering engagement with the aggressors.

This sobering situation starkly underscored what General Creech came to call "go-low disease," motivated by his concern that the emphasis on low-altitude ingress was not only causing a needlessly high accident rate in peacetime training, but also was jeopardizing aircrew survivability and future flexibility in actual combat while, at the same time, constraining TAC's appreciation of the equipment needed to perform the ground-attack mission more effectively. In response, Creech insisted on new tactics aimed at making defense rollback the first order of business. Since the most lethal Soviet SAMs could not be successfully underflown within the heart of their engagement envelopes in any event, the emphasis instead swung to developing equipment and tactics that would enable the opening of a medium-altitude window, even as aircrew proficiency at low-level operations was maintained as a fallback measure. The new focus concentrated on sanitizing the air defense environment by taking out or neutralizing enemy SAMs as a first priority, so that attacking aircraft could operate more safely as soon as possible at higher altitudes beyond the lethal reach of AAA.

At the same time, motivated by a determination to insert not just greater tactical but also operational and strategic realism into Red Flag, General Creech eliminated the initial "core squadron" mission planning practice and instead put TAC's air division commanders in charge of scenarios on a rotating basis. At the same time, there emerged a heightened emphasis on acquiring the needed equipment that would render medium altitude tactics both possible in principle and also effective. This included acquiring the EF-111 electronic jammer, advanced antiradiation missiles, and electro-optical and laser-guided precision munitions. The new effort also included an expenditure of more than $600 million to increase the scope and realism of the enemy threat simulators fielded throughout the Nellis Range Complex, as well as range measurement instrumentation and the Red Flag Mission Debriefing System (RFMDS), which allowed a detailed real-time monitoring and subsequent reconstruction of each event performed by each aircraft operating within the exercise's air space. At the same time, new capabilities and tactics for operating at night were pushed hard and ultimately validated at Red Flag. Thanks to that, the character of Red Flag shifted notably toward something more closely approximating realistic large-force employment against an enemy whose defenses would eventually be degraded in actual combat. The result was more real-world training realism, as opposed to the false realism of an impenetrable enemy defense, which was finally understood in hindsight to have produced more negative than positive training.

Getting Serious About Electronic Warfare

Closely connected to this stress on greater realism and greater emphasis on enemy air defense suppression was a mounting concern over the need to introduce the complexities of electronic combat into peacetime tactical training, especially those connected with coping effectively in a heavy communications jamming environment. Both during his previous assignment as the commander of the USAF's Electronic Systems Division and later as TAC commander, General Creech figured prominently in this effort to integrate a serious program of offensive and defensive electronic combat into the Air Force's training repertoire. In 1981, he initiated Green Flag, a Red Flag-like exercise conducted biennially at Nellis with special emphasis on electronic warfare and SAM suppression.

During the first Green Flag, Creech directed that communications jamming be turned on at the outset and left on throughout the operation just as the Soviets would do in actual combat. As a result, 72 percent of the training sorties flown were ineffective. That ended once and for all the assumption that one could overcome enemy jamming efforts merely by manually changing radio frequencies. At the outset of Red Flag, a procedure was employed in which once simulated comm-jamming began, the mission commander would call "chattermark" over the common strike frequency and all aircrews would immediately switch to an alternate prebriefed frequency. It soon became apparent that the fast-hopping enemy jammers would quickly identify the new frequency and begin jamming it as well. Indeed, once serious comm-jamming began in Green Flag, aircrews discovered they could not even transmit a complete "chattermark" radio call before being hopelessly cut out. That led to a realization that the "chattermark" procedure would never work in combat. It also confirmed the need for antijam radios and helped to justify funding for more serious preparations for electronic warfare, including a program called Compass Call, which entailed the installation of state-of-the-art jammers on C-130s to disrupt enemy voice communications.

TAC Turnaround

Along with the major advances in aircrew training and proficiency outlined above, a largely unsung but nonetheless groundbreaking parallel improvement also took place in the organizational efficiency of TAC during the late 1970s and early 1980s under General Creech's tutelage as commander. Earlier in the 1970s, upward of half of TAC's $25 billion inventory of aircraft were not mission-ready at any given time, and as many as 200 of its 3,800 aircraft were classified as "hangar queens"—grounded for three weeks or more due to a lack of maintenance or needed parts. Moreover, pilots who required a minimum of 20 hours of flying time a month to remain operationally ready were getting only half that amount in most cases.

At the same time, TAC during the early years after Vietnam suffered high maintenance inefficiencies and an unacceptably high accident rate that was partly occasioned by them. Unwilling to reduce fielded squadron strength to absorb the deep funding cuts of the Carter administration, the Air Force leadership accommodated this financial crunch instead by raiding its operations and maintenance accounts. All of this was heavily driven by the top-down management style that had come to afflict the entire US defense establishment as a result of Defense Secretary Robert McNamara's prior imposition of a dogma of centralization from the business world which, by the end of the Vietnam war, had pervaded almost all walks of American military life.

Among the many pernicious results of this affliction was a mounting lapse in integrity at the operational level, in which small lies about unit performance became ever larger sins of self-deception which ultimately undermined both mission readiness and safety. Driven by a perceived need to worship statistics for their own sake rather than the underlying facts they were supposed to represent and by a bureaucracy which insisted on hearing the "right" answers irrespective of reality, USAF aircrews would falsify their mission reports as deemed necessary to show that they had performed events such as inflight refuelings and weapons deliveries which they had in fact not conducted. Thanks to the same felt compulsion, unit supervisors would record takeoffs which had been delayed by maintenance as "on time" and assign aircraft to the flight schedule which had not been properly released by maintenance control. In sum, bureaucratic gridlock and an overlay of regulations and statistical imperatives, aggravated by diminished funds, had come to stifle morale and to discourage initiative and innovation at the command's grass-roots level.

With the strong backing of Air Force Chief of Staff General David Jones, Creech quickly sized up the situation and proceeded to invert the traditional top-down centralization of TAC by imposing a strict bottom-up approach to the organization and management of his command, in the process forcing authority and responsibility down to its very lowest reaches. At the same time, he introduced a radically new and different tone by replacing the former pattern of leadership intimidation and bluster with what he called "reasoned command." The new watchword became management through motivation rather than regulation, on the premise that professionals will willingly assume greater responsibility when they are treated with dignity and given a sense of personal ownership of their contribution to the larger whole.

Creech's leadership philosophy, while clearly cognizant of the fact that military organizations are not democracies, was in effect a common-sense variation on the golden rule. It was based on a recognition that loyalty was a two-way street and on the premise that if a commander always looked up to those at the front, he would never talk down to them. It was profoundly intolerant of centocratic practices and recognized that an organization can only be as successful as those at the bottom are willing to make it. Toward that end, Creech emphasized focusing more on the product than on the process. He sought with determination to minimize excess regulation, which he believed merely depressed the spirit and stifled motivation. He also sought to replace inhibitions on communication with full openness, and he shifted his headquarters function from restricting to facilitating. Above all, he constantly stressed that there were no poor units, only poor leaders.

In a related vein, Creech insisted that a mistake was not a crime and a crime was not a mistake, and he incessantly played up the importance of honoring the difference between the two in meting out discipline for mishaps and lesser oversights. His abiding goal was to infuse the system with trust and respect so that coherence and control might be maintained through incentive rather than through top-down authoritarianism. At bottom, he sought to instill throughout the ranks an appreciation of the crucial difference between quality *control* and quality *creation* and to focus predominantly on the latter, which demanded both different language and a different mindset. To achieve it, he strove to inhibit excess micromanagement of inputs from above. He also spotlighted pride as the fuel of all human accomplishment, a quality that needed creating and sustaining by empowering those at the working level to show initiative, while providing for responsibility and accountability at every level.

As Creech later explained it, "the villain wasn't any particular person, but the whole system." What he did by way of response to that affliction was to dismantle the former centralized management regime at the wing level and to empower subordinate squadrons to do their own aircraft maintenance, in the process giving each squadron a sense of identity, spirit, and purpose, along with a corporate stake in the fruits of its efforts. By systematically pushing decisions down to the level of those front-line supervisors who actually carried them out, the risk of poor decisions was sharply reduced. Creech personally played a lead role in selecting, mentoring, and grooming those at the working level who showed the greatest promise for future leadership, motivated by his credo that the cardinal imperative of a leader is to produce more leaders. His four simple "pass/fail" standards of conduct expected of all subordinate TAC leaders entailed a staunch refusal to countenance any manifestations of lying, displays of temper, abuse of position, or lapses in integrity.

The payoff of this turnaround in the TAC culture soon became widely apparent. Time came to be utilized more efficiently, quality in all domains of command activity went up, and excellence became a TAC-wide fixation. Units became competitive in all major areas of endeavor, particularly in maintenance delivery and flight operations. Unit commanders were encouraged to fly more often and to lead from the front. All of this generated measurable improvements in all major categories of performance with no more aircraft, personnel, or money than TAC had when Creech first assumed command. He narrowed the gap in trust between TAC's leaders and led, installed a system based on mutual respect and mutual support, sought fervently to eliminate the "not invented here" syndrome, and instilled a quality mindset at every level, basing the product (TAC's organizational efficiency and mission readiness) on persuasion rather than ex cathedra orders. The ensuing effect of reducing the number of TAC's aircraft that were down for maintenance at any given moment by three-fourths yielded an inventory availability and increase in combat capability from existing assets that would have cost more than $12 billion had they been purchased anew.

These reforms eventually permeated all elements of TAC down to the lowest front-line operators, in the process fundamentally changing their former roles, relationships, and responsibilities by enhancing the creativity and commitment of those who ultimately determined the command's success. The impact of the reforms on TAC's morale quotient was palpable. Not long after they were set in motion, TAC's first-term reenlistment rate increased by 136 percent, a resounding vote of approval for the new, decentralized, and team-based approach to management. Other early returns included a substantial increase in available flying hours for aircrews, better quality of aircraft maintenance, and a sharp increase in TAC's overall mission readiness rates, all *before* the Reagan-era defense dollars finally became available in 1983— from the 1981 budget year appropriation—to enhance the nation's military strength even further.

Impact of "Robusting"

In a significant departure from previous resource management practice which he called unit "robusting," Creech dramatically increased the combat capability of TAC's wings simply by applying a new organizing principle. Instead of sharing shortages across all three squadrons in a given wing, as had been the previous practice, Creech took the existing assets of a wing (both assigned aircrews and aircraft) and built up two of the three squadrons to full strength rather than spreading the pain equally among all three. Hitherto, nearly every TAC squadron command-wide found itself two to three aircraft short of its authorized strength, and new units were being filled at the expense of existing units thanks to a command determination to adhere to assigned schedules irrespective of procurement shortfalls.

Indeed, in the summer of 1978, because of equipment shortages and the fact that some combat-coded units were being devoted full-time to aircrew replacement training, TAC had only 33 of 45 authorized squadrons actually assigned to line wings. There were also significant aircrew and spares shortages against approved mobilization and deployment plans, even though those plans were being honored in name all the same. As a result, the actual aggregate combat capability of TAC's wings was diluted, since their squadrons were assigned wartime responsibilities but not the needed resources to meet them. That, in turn, yielded a surfeit of opportunities for excuse-making and self-deception by masking rather than spotlighting real shortages. It further had the effect of creating units with a nominally good mission-readiness image, yet which would have been forced to reorganize to meet their wartime commitments. Lapses in integrity were only a short step away, with predictable consequences in sagging morale.

The corrective measures instituted by General Creech in 1978, with full Air Staff support, established explicit criteria for "robust" units, namely, those that were manned and equipped at a level that made them ready to meet their wartime tasking. He did this by filling out one squadron in a wing first, and then the second, and leaving only the third squadron to shoulder the wing's shortages. He further deferred activating new units until existing ones were filled to their authorized levels. The advantages of this new approach included greater honesty in unit status reporting by highlighting rather than hiding shortages and by sharing a wing's strengths rather than its weaknesses. Robusting made TAC's wings better organized for prompt deployment to meet wartime obligations by preventing the suboptimization of key assets and by putting pressure where it belonged, namely, on resource suppliers so that unit tasking would be more in line with actually available resources.

In all, these reforms lent a sharper focus to authority and accountability and got unit-level peer pressure working in positive rather than negative directions. They also drove authority, accountability, and a sense of ownership to the lowest possible levels throughout TAC, giving everyone in the system both pride of involvement and a personal stake in the product. In short order, thanks to the breaking down of functional fiefdoms sealed up within functional walls and the replacement of nonintegrated job-oriented functions with integrated, product-oriented teams, TAC went from a vertically- to a horizontally-organized command. Each squadron became responsible for its own 24 assigned aircraft, with all disciplines working together in small teams within the squadron to get the job done. Along the way, paperwork was reduced by 65 percent, and the average time required to deliver a needed part was lowered from three and a half hours to eight minutes. The net result was a genuine personalizing of a once-impersonal system, as well as a doubling of the number of peacetime training sorties flown during a given training period with no increase in operating cost. As a result of these initiatives, by the time Creech left TAC and retired from the Air Force in 1984, 85 percent of TAC's aircraft were rated mission-capable, and fighters were averaging 21 sorties a month, more than twice the level that prevailed when he assumed command in 1978.

In an enduring legacy of the TAC turnaround, this transformation in management style and organizational efficiency instituted during Creech's tenure later swept the rest of the Air Force. The team-based, decentralized approach to management flowed from the premise that desired accomplishments are achieved by individuals and small collectives working as teams.

Payoff in Improved Safety

Abetted to a substantial degree by the management reforms described above, a sharp decline in the USAF's aircraft accident rate between 1975 and the early 1990s bore out a point long argued by Airmen the world over that training aggressively to the limits dictated by the needs of operational realism produces not only more combat-capable pilots, but also safer pilots.

From 1975 to 1995, the annual number of Air Force Class A mishaps (those accidents involving a fatality, the loss of an aircraft, or equipment damage amounting to $1 million or more) dropped from 99 to 32. The mishap rate per 100,000 flying hours between 1975 and 1995 fell from 2.8 to 1.4. In the most critical category of fighter aviation, the incidence of mishaps fell by 61.5 percent since 1975, with the curve flattening out, for the most part, by 1985 even as the number and intensity of sorties increased. In contrast, in routine peacetime training during the 1950s and early 1960s, TAC routinely lost upward of 14.6 aircraft per 100,000 flying hrs. This indicated that in addition to greatly enhanced combat readiness, the new training approach introduced after Vietnam was, along with more reliable aircraft and a heightened stress on safety, also producing a windfall dividend in reduced aircraft losses that had not even been anticipated, let alone planned for, until the Creech reforms took root throughout TAC. The accident rate in TAC subsequently dropped from one in every 13,000 flying hours to one in every 50,000 hours, even as the peacetime low-level training limit was lowered from 500 ft to 100 ft, increasing the risk factor as a buy-in cost for greater training realism, yet placing trust in the ability of greater latitude and professionalism at the working level to compensate for excess rule-mongering and stifling top-down micromanagement.

Case Study: General Ryan and Creation of the AEF

Introduction.

General Michael E. Ryan was the Chief of Staff of the US Air Force from October 1997 to September 2001 where he served as the senior uniformed Air Force officer responsible for organizing, training, and equipping the 700,000 active-duty, Guard, Reserve and civilian forces serving in the United States and overseas.

General Ryan participated in two interviews, the first of which investigated his views on leadership; the second session related those views to his experience while launching the expeditionary air force (EAF) and aerospace expeditionary force (AEF) constructs. Other interviews with his contemporaries, such as the present Air Force Chief of Staff, General John P. Jumper, and the present Chairman of the Joint Chiefs of Staff, General Richard B. Myers, provided keen insights into General Ryan's leadership skills and the application of those skills during the implementation of EAF/AEF.

It goes nearly without saying that a man of General Ryan's experience has a deeply considered view of leadership. Surprising though is the simplicity with which he expressed those views. Throughout the interview sessions he frequently returned to two themes: integrity and three simple keys: have a plan, put someone in charge, and follow up. Certainly there were other leadership techniques he thought important, such as career broadening and cross-pollination, all of which will appear in the following discussion; however, General Ryan's thoughts on leadership were very direct and succinct—clearly the distillation of many years of experience. Some things work; some do not, and General Ryan knows the difference.

Integrity

Throughout his career, General Ryan has observed that the bond of trust cemented by integrity is one of the keys to success. Whether it is flying on the wing or taking a briefing to the Pentagon, "ordinary people do extraordinary things…it has to do with trusting" both up and down the chain of command. One other important factor that builds trust is common sense, a virtue General Ryan holds in high regard. "Common sense is *the* leadership trait along with integrity." One trusts a subordinate or superior because he or she has common sense and integrity. As we will see later in the AEF discussion, General Ryan was not reluctant to put someone in a position based on the subordinate's integrity and common sense even though the subordinate's technical background seemed at odds with the job. "When you put them in a job they know little about, they learn faster. You come back in four months, and they're brilliant. People rise to new challenges."

General Ryan learned this technique of cross-pollination early in his career. As a young F-4D pilot at Holloman AFB, New Mexico, his squadron commander thrust him without notice and virtually no preparation into being the squadron maintenance officer—this in the face of an impending remake of a failed operational readiness inspection (ORI). Although he thought it was "stupid" at the time, he now reflects, "We passed with flying colors…and…even though I never picked up the Air Force specialty code (AFSC) I always treasured the challenge and experience."

A second occasion of career broadening happened when he arrived as a major on the staff at Tactical Air Command. Slated for a flying job on the inspector general's staff, a general officer diverted Ryan to the TAC/XP (plans) staff—a non-flying position. "I hated it," he says, "but it was the best thing that ever happened to me." Mentoring became a reality to the young major, and he carried the technique into his years at flag rank. "Somewhere in the Air Force we must make [broadening] part of our growing up process. You can't just let it happen…this broadening business has got to be…institutionalized." Broadening and mentoring are important, but so is the responsibility that falls to the protégé. The mentor may place the officer in a challenging position, but it is up to the protégé to succeed—often in the hardest and most demanding jobs in the Air Force.

Thus role modeling and mentoring, as well as the tenets of leadership and followership, heavily seasoned with common sense, flavored General Ryan's leadership style as he served the Air Force and his nation. As he observed a few months after becoming the chief, "I have heard the lament that, 'the Air Force is not what it used to be during the Cold War,' and I must tell you that is absolutely true; this ain't our fathers' Air Force."

Have a Plan

After his induction as chief of staff in October 1997, General Ryan worked to bring the USAF into line with the realities of the post–Cold War era, what General John P. Jumper called a "transition out of the Cold War mindset." Just as striking was the reduction in force structure. After 1990, fighter wings fell from 24 active-duty and 12 reserve wings to 13 active-duty and 7 reserve wings. Bombers declined by 50 percent, tankers by 40 percent and cargo/transports by 25 percent. It was the smallest Air Force since its founding in 1947.

Yet, the nation's strategy of selective engagement dictated that the Air Force be ready to fight and win two nearly simultaneous major theater wars, while maintaining its commitments to a growing string of seemingly permanent small-scale contingencies, which required the opening of numerous expeditionary bases. From 1992 to 1996 the US armed forces continued three major operations and initiated seven more. In Ryan's words, "…we had done Desert Storm, Somalia, Bosnia, and it looked like a never-ending chain of these things was going to occur…it didn't look like there was any end." The mismatch between resources and requirements was forcing the men and women of the USAF into a lifestyle characterized by high operations and personnel tempo at the expense of family life. Drops in Air Force retention rates and recruitment indicated that the situation, if allowed to go unchecked, could reach serious proportions. General Ryan observed, "We were going to get picked to pieces."

Acting quickly and decisively, the new chief implemented several initiatives to relieve the stress of operations tempo on his force. Shortly after his induction, he issued a special Notice to Airmen (NOTAM) on retention that announced:

- A 5 percent reduction in USAF and joint training exercises for the next two years,
- A 15 percent cut in Service personnel supporting Chairman of the Joint Chiefs of Staff exercises,
- Termination of Quality Air Force Assessments,
- A 10 percent reduction of the length of inspections and the number of inspectors used in ORIs for FY 98 and an additional 20 percent reduction for FY 99,
- Combining inspections with real-world deployments, and
  improved training for those deployed to Southwest Asia.

Beyond the burdens of operations tempo and shrinking force structure were other questions that had been nagging and "frustrating" General Ryan for years: Who are we? How do we express who we are? He saw the Navy express itself in terms of carrier battle groups; the Marines in terms of Marine expeditionary forces; the Army in terms of a corps. Yet when someone called for the Air Force, "Sometimes four guys showed up. The space guy. The Air Combat Command (ACC) guy. The air mobility (AMC) guy. And then some special operator…we weren't unified, and…we had no unifying concept or construct in the Air Force." Without identity the second question is even more difficult to answer especially in light of the collapse of the Soviet Union. The Air Force had nothing to structure against, "We didn't have a force structuring mechanism."

Nearly every treatise on leadership, especially senior leadership, ascribes vision as a characteristic of great leaders. General Ryan is more pragmatic, "Vision is an overused word. I boiled it down to three things. You must have a plan. You must have somebody in charge. And you must follow up." Further, he learned early as a new flag officer to think in terms of the future, oftentimes the distant future. "I don't think I thought that far ahead…when I was a [wing] commander. It was a world of…now…or maybe a year or two ahead. I was very naïve about how important [the long term] was. For me it was a matter of dragging myself up to think more into the future than I had done as a colonel. To start conceptualizing about what we ought to be doing."

The concept in General Ryan's sights was an expeditionary force, defined to parallel the Navy's battle group and the Marines' expeditionary force—a force that would alleviate if not solve the Air Force's internal problems, but also one that would express to the nation and the force structure gurus what the Air Force was and how it operated. To some the task was daunting, but his predecessor and many others had already poured the footings of the concept by beginning development of expeditionary forces for employment in Southwest Asia and elsewhere. Certainly General Ryan was not hesitant to step out; as General Richard B. Myers said, "He was not afraid to set course on a new heading."

General Ryan's new course for the Air Force was the Expeditionary Aerospace Force (EAF)—a new way of doing business that improved predictability and stability in personnel assignments and furnished the Service with a powerful management tool to more efficiently align its assets with needs of the war fighters—the regional combatant commanders (RCC). To some this was vision; to General Ryan it was step one: You must have a plan.

Put Someone in Charge

The notion of a quick-reaction airpower force, consisting of several types of planes, had been a thread woven through Air Force thinking for at least 40 years. Then in October 1995, the Navy's rotational schedule called for USS *Independence* to withdraw from US Central Command's (USCENTCOM) area, but the Navy was unable to replace *Independence* for six weeks. Thus the task of creating a "carrier gap filler" fell squarely on the USAF Ninth Air Force's commander. General Ryan recalled, "We started as a carrier gap filler...a kind of carrier equivalency that the [combatant commander] would [understand]." These carrier gap filler forces provided the test bed for what General Ryan and others would ultimately view as the solution to many problems—AEF packages that would "operationalize" the EAF culture already existent but neither clearly defined nor cogently communicated. Without question the carrier gap filler expeditionary forces produced a belief throughout the Service that the USAF must fundamentally change its operations, structure, and culture to adapt to the different and difficult conditions of the post-Cold War era. General Ryan recalls a meeting where he tasked them to apply the expeditionary notion to the Air Force at large. They first came back with a rotation of combat forces. He told them, "No, I want intelligence, surveillance, and reconnaissance (ISR) in there, and support forces. I want C-21s and C-130s. I want this to be pieces of all the Air Force that goes forward."

At Corona South in February 1998, General Ryan personally presented his plan to other senior service officers on the first day of the conference. He directed nine tasks each with a responsible office. Generally grouped, the tasks can be seen as: have a plan, put someone in charge, and follow up. In this case the follow up was to be the next and succeeding Coronas.

Figure D.1. Life Cycle of an AEF

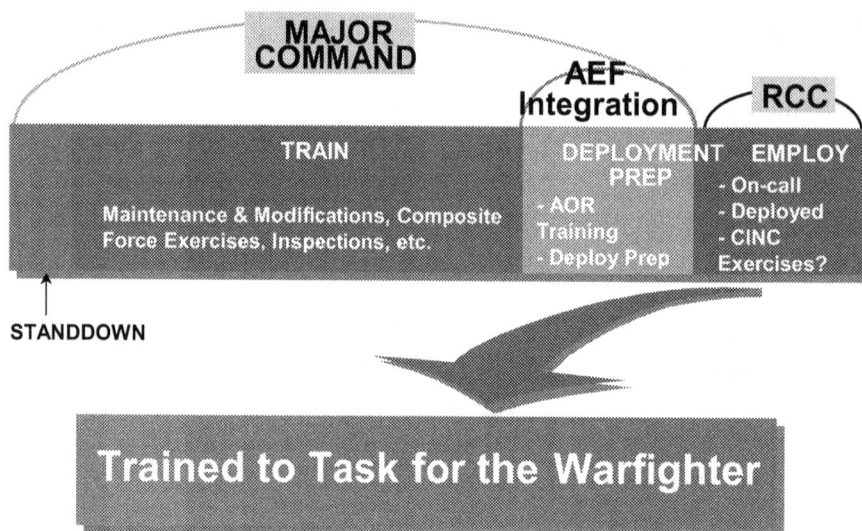

Of course the development of AEFs went through many iterations. The final solution, though, was 10 AEFs set up on regular rotations, 2 at a time. Figure D.1 above describes the flow of preparation, deployment, and reconstitution for the AEFs.

Of interest to this case study, however, are the inputs and directives General Ryan made based on his leadership style and his sense of responsibility to the people and the mission of the Air Force. As General Jumper put it, "there's no limit to what you can do if you don't care who gets the credit...He [Ryan] took something that was already out there, saw the good in it and found a way to apply it to the whole Air Force."

First, General Ryan clearly was interested in the impact of AEF on his people. In his words, "This was about family. If the family is disgruntled because the [Service] member has no predictability in his life, they're going to walk. They have 51 percent of the vote." The regular schedule of AEF rotations aimed to give the families predictability in their lives as well as to provide the combatant commander with a superb air component. Apportioning combat forces was straightforward because the people were tied, in most cases, to a particular weapons system, but dividing support forces into the AEF was no easy task. Despite the draw down of permanent overseas bases, the Air Force had opened many small deployment bases in order to meet continuous taskings. In General Ryan's words, "we opened a gazillion of these things, and...we kept them open. We were taking the manpower sized for our permanent bases and spreading it across these expeditionary bases. We were shorting ourselves at home...TDYing ourselves to death."

Over a year period, the chief called each functional area manager to his office and required the officer to account for every billet's availability to join the AEF construct—this from bases that normally did not deploy—in order to insure a fair and equitable distribution of the deployment workload. His goal was to get the base of AEF manpower up to 200,000, or about 60 percent of the force. He said, "I think we're just about there, which is good. When you have that many you don't have exceptions to the rules, [and] you get a rotation going where people aren't deployed all the time."

There is more to combat capability than fairness, and another of the former chief's leadership beliefs is that Air Force people are tribal; they want to be part of the group, and once a part of the group they work best by staying together. During a visit to a desert air base, he visited the fire department. While talking to them, he asked the names of their home bases. No two of them were from the same base. Firefighters rely on voice recognition and oftentimes with hand signals. They train together, and in General Ryan's view they should deploy and fight fires together. He also sees another side to the tribal coin: the issue of taking care of the families at home. If one member of a team is gone, the family copes alone. If the whole team is deployed, the families enjoy a synergy of support from within as well as from without. "So there is an element of effectiveness on the line and one on the home front...you go in teams." He continued, "You cannot measure how important it is for people to understand that their peers understand and appreciate their contribution."

In one iteration of sizing the AEFs, they were different in the numbers and types of aircraft in the force. General Ryan had the experience as a force employer when he served as the Sixteenth Air Force commander, and he had the experience as a force provider when he commanded United States Air Forces in Europe (USAFE). Having walked both sides of the street, he had a deep appreciation for the fact that no commander wants to think he is not getting the best. Thus, as chief, General Ryan insisted that all the AEFs have nearly the same size and abilities to accomplish the combatant commander's missions. There would be no "A-team" and "B-team" in his method of operation with the AEF.

Holding true to his second tenet of putting some one in charge, he rejected one staff proposal with the comment, "Where's the leadership," and directed the staff to plan the AEFs around a leadership core consisting of a general officer with a staff. Two things prompted his direction. He saw the Air Force as having firmly established operational level leadership plans for most operations—in Korea, and the Gulf, as well as others. However, he did not see the same leadership scheme at the expeditionary base level. Thus, he required "lead" wings in the AEF rotations on both the mobility side and the combat air forces (CAF) side so there would be no *ad hoc* leadership schemes kluged at a deployment base on short notice. "You must have somebody in charge—the wing commander of that lead wing. He's going to be in charge of opening up that base."

Follow Up

Sometimes following up is picking the right person to continue an effort already underway. Most would agree that one of the senior leader's most important tasks, often rife with risk, is the selection of subordinates—who is the right person for the job? General Ryan took a personal interest in selecting the general officer that would lead the new Directorate of EAF Implementation "to provide policy guidance and provide oversight as the HQ USAF point of contact for all EAF implementation activities." He selected a major general because "he didn't know anything about it!" Ryan wanted somebody who had not been close to the problem, and although the general he selected had a broad experience in bombers, missiles, and space, he had not been deeply involved in the many versions of EAF/AEF. As Ryan said, "[This officer] was buffered from it, [and] I wanted somebody who could ask all the questions. So I said come to Washington and your qualification is that you don't know, so you'll ask."

This cross-pollination technique is one General Ryan used frequently during his tenure as the chief of staff. He sent a navigator to command a numbered air force—the first officer of that rating to hold the job. He sent an airlifter to command the Pacific Air Forces, and a special operations specialist to be the vice commander of the United States Air Forces in Europe, predominately fighter commands. "I wanted a mix of experience," he said, "because they were growing up in these stovepipes, and that didn't help the Air Force's broad view." He continued, "I was moving general officers around in non-traditional places because I thought general officers ought to be general."

General Ryan did not see an advantage in locating the final EAF center in a non-traditional location. In fact he directed it to be sited at Langley AFB, the home of Air Combat Command (ACC), a command deeply seasoned in forming deployment packages by unit type codes (UTC). The success of the EAF hinged on properly allocating support forces, and "we had never really done that [correctly] before. So, this was part three of his credo to have a functioning organization that can "follow up." He said, "I told [the ACC commander] you're doing this not just for the CAF but for all the Air Force. We're putting it at Langley because you normally do the rotations for the rest of the CAF. You're going to do it for the CAF, mobility air forces (MAF), AETC, and the logistics command and everybody else." Clearly General Ryan made some decisions based on traditional practices, yet others were counter-intuitive—selecting a solid thinker not because he had specific expertise but because he was "a good man" who would question traditional methods.

Coordinating an effort of the magnitude of the EAF implementation and its attendant AEF packages can be risky business especially at the senior officer level, perhaps more so when the push is coming from inside the Pentagon out to the field commands. After thoroughly analyzing and comparing his inventory of combat aircraft to the current requirements of the combatant commanders, the ACC commander concluded that, in the short term, he could not field 10 AEFs of roughly equal capabilities. Although he hoped to reach the ten-AEF goal in the future, he directed his staff to prepare a new structure for the first AEF rotation cycle. He proposed a construct of nine AEFs that would deploy only to existing contingencies, such as the no-fly zones, plus one on-call AEF that would supply pop-up and new contingency deployments. The alternative proposal was called the 9+1 formula. The reasoning behind the 9+1 formula reflected existing shortfalls that could only be cured by additional resources. The most critical shortage was the lack of aircraft and pilots equipped and trained for suppression of enemy air defenses. The split-operations capability that the original AEF structure—known as the stacked 10—required of the existing squadrons added a further difficulty. Squadrons would not only have to supply a force for a scheduled deployment, but they might well be called upon to supply a second forward force to respond to a new crisis. The strain of maintaining just one forward location and the home station had already stretched the Service's resources. ACC believed its units could not routinely support two forward locations.

General Ryan had never been completely happy with the 9+1, in part because it seemed to contradict the public promises he had made to his service earlier. General Ryan met with the ACC commander to discuss EAF matters, and they agreed to abandon the 9+1 in favor of ten AEFs backed up by a dedicated on-call force. General Ryan recalled, "I looked at it hard. I [was] trying to work the theory of it for future force structuring, and they were trying to work the reality of today's needs. So we just worked our way through it."

Another thorny coordination issue was the combatant commanders, who would receive the new AEF packages. General Ryan briefed the AEFs to the combatant commanders several times, both at their conferences and to them individually, and he had the combatant commanders' air component commanders do the same. "...the [combatant commander] said, 'Wait, let me get this straight, you're going to take all my deployed forces and change them out over a 2-week period?' Yeah. I said, 'These people know where they're going; trained to do it...they come in teams. They'll hit the ground running. You'll love it!'"

What General Ryan anticipated as the hardest sell, the Pacific Command, turned out to be easier than he thought. The Pacific combatant commander was a naval man thus very experienced with rotational forces. "...the Navy said, 'You're getting to look like us.' I said, 'Exactly. We stole it from you. You...and the Marines, you've got it right.'" According to General Myers, one of General Ryan's strongest leadership traits was his "collaborative" style where everyone involved had a voice, a point that comes out clearly in the above examples. Where other leaders might have dug in their heels and reacted with emotion or hostility, General Ryan used his persuasive skills. General Myers said, "[His collaborative style] served him very well...It made people willing to tell him what was good about an idea; what was bad."

Continuing on his tenet of following up, General Ryan made sure the Air Force clearly defined how the AEFs would recover, or reconstitute, after they returned from a rotation. Figure D.1, "Life Cycle of an AEF," shows a stand down period at the beginning of the cycle. Another graphic that the general frequently used to argue the EAF case more clearly expresses his philosophy of giving man and machine time to recover. Figure D.2 shows the build-up from two AEFs deployed through the build to a major theater war and the subsequent reconstitution.

EAF Across the Spectrum

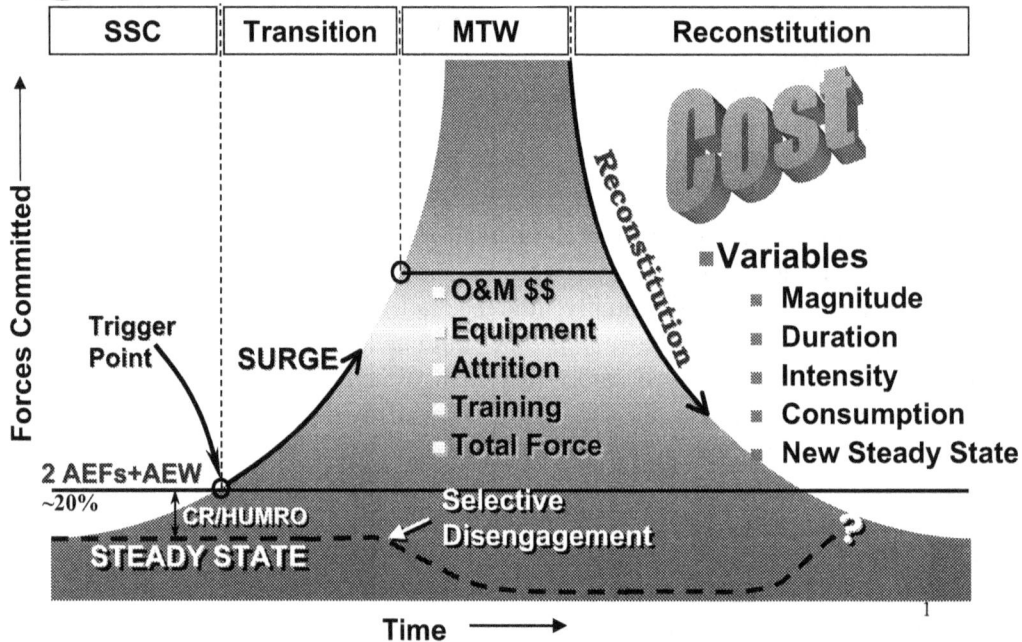

| SSC | Transition | MTW | Reconstitution |

Forces Committed

Cost

Trigger Point

SURGE

- O&M $$
- Equipment
- Attrition
- Training
- Total Force

Reconstitution

- **Variables**
  - Magnitude
  - Duration
  - Intensity
  - Consumption
  - New Steady State

2 AEFs+AEW
~20%

CR/HUMRO

STEADY STATE

Selective Disengagement

?

Time →

Figure D.2. EAF Across the Spectrum

During the critical final stages of planning to implement the EAF, the Air Force was engaged in Operation ALLIED FORCE, an effort that required a level of effort close to that of a major theater war—especially for "low density/high demand" systems. The implementation date for the EAF was only a few short months away on 1 October 1999. Many wanted to delay implementation to reconstitute forces returning from ALLIED FORCE, but General Ryan said, "…no. Use this to recover and implement the AEF around the recovery, because AEF is also a recovery tool. [To] those who said…delay…I said, 'Hell, no. We ought to accelerate it!'" This despite the fact that planning for the implementation was behind schedule because of the realities of combat in ALLIED FORCE. Most observers agree that the Air Force had reached about an 80 percent solution, yet General Ryan and the Secretary of the Air Force decided to give EAF the go-ahead. He recalled, "I told them 80 percent was good enough, and we'd tell the field it was 80 percent. In fact, [any decision] will always be something less than 100 percent. If we had waited any longer, we would have lost the rationale for reconstitution."

The AEF structure began operations on schedule on 1 October 1999 when AEFs 1 and 2 deployed. Seventy-four percent of this force was sourced to UTCs, a figure that would rise to 94 percent when AEFs 5 and 6 deployed on 1 March 2000. EAF was an idea whose time had come, and as one historian has noted, "Few changes introduced by an Air Force chief of staff have flowed as smoothly through the corporate process as did the EAF."

Conclusion

According to his successor, General Jumper, General Ryan's three strongest leadership skills are laced throughout the implementation of AEF. First was his selflessness: "Mike Ryan is the one who inspired [people] and then backed them up completely." Second, his sense of humor carried the day when times were hard. General Jumper recalled, "I saw [his good humor]…up close and personal, and [it] is a trait we should all try to emulate." Third, he had an "uncanny ability to detach himself from the problem and take a very impersonal, analytical look…without the emotion attached."

GLOSSARY

Abbreviations and Acronyms

AAA	antiaircraft artillery
ACC	Air Combat Command
AEF	air and space expeditionary force
AETC	Air Education and Training Command
AFB	Air Force Base
AFDC	Air Force Doctrine Center
AFDD	Air Force doctrine document
AFSC	United States Air Force specialty code
AOC	air and space operations center
CAF	combat air forces
CMSAF	Chief Master Sergeant of the Air Force
COMAFFOR	commander, Air Force forces
CSAF	Chief of Staff, United States Air Force
FDS	foundational doctrine statements
ISR	intelligence, surveillance, and reconnaissance
JFACC	joint force air component commander [JP 1-02]; joint force air and space component commander {USAF}
MAF	mobility air forces
MAJCOM	major command
NAF	numbered air force
NCO	noncommissioned officer
NOTAM	notice to Airmen
OODA	observe, orient, decide, act
ORI	operational readiness inspection
ORM	operational risk management
RFMDS	Red Flag Mission Debriefing System
SAC	Strategic Air Command
SAM	surface-to-air missile
TAF	tactical air forces
TDY	temporary duty
US	United States
USAF	United States Air Force

USCENTCOM	United States Central Command
UTC	unit type code
WW II	World War II

Definitions

air and space expeditionary force. An organizational structure to provide forces and support on a rotational, and thus relatively more predictable basis. They are composed of force packages of capabilities that provide rapid and responsive air and space power. Also called AEF. (AFDD 1)

air and space power. The use of lethal and nonlethal means by air and space forces to achieve strategic, operational, and tactical objectives. (AFDD 1)

Airman. Any US Air Force member (officer or enlisted, active, reserve, or guard, along with Department of the Air Force civilians) who supports and defends the US Constitution and serves our country. Air Force Airmen are those people who formally belong to the US Air Force and employ or support some aspect of the US Air Force's air and space power capabilities. The term Airman is often used in a very narrow sense to mean pilot. An Airman is any person who understands and appreciates the full range of air and space power capabilities and can employ or support some aspect of air and space power capabilities. (AFDD 1-1)

continuation training. Training to maintain basic skill proficiency or improve the capability of individuals to perform the unit mission. (AFDD 1-1)

core values. A statement of those institutional values and principles of conduct that provide the moral framework within which military action takes place. (AFDD 1-1)

doctrine. Fundamental principles by which the military forces or elements thereof guide their actions in support of national objectives. It is authoritative but requires judgment in application. (JP 1–02)

education. Instruction and study focused on creative problem solving that does not provide predictable outcomes. Education encompasses a broader flow of information to the student and encourages exploration into unknown areas and creative problem solving. (AFDD 1-1)

force development. A series of experiences and challenges, combined with education and training opportunities, that is directed at producing Airmen who possess the requisite skills, knowledge, experience, and motivation to lead and execute the full spectrum of Air Force missions. (AFDD 1-1)

joint force air component commander. The commander within a unified command, subordinate unified command, or joint task force responsible to the establishing commander for making recommendations on the proper employment of assigned, attached, and/or made available for tasking air forces; planning and coordinating air operations; or accomplishing such operational missions as may be assigned. The joint force air component commander is given the authority necessary to accomplish missions and tasks assigned by the establishing commander. Also called **JFACC**. See also **joint force commander**. (JP 1-02) [The **joint force air and space component commander** (JFACC) uses the joint air and space operations center to command and control the integrated air and space effort to meet the joint force commander's objectives. This title emphasizes the Air Force position that air power and space power together create effects that cannot be achieved through air or space power alone.] [AFDD 2] {Words in brackets apply only to the Air Force and are offered for clarity.}

joint publication. A publication containing joint doctrine and/or joint tactics, techniques, and procedures that involves the employment of forces prepared under the cognizance of Joint Staff directorates and applicable to the Military Departments, combatant commands, and other authorized agencies. It is approved by the Chairman of the Joint Chiefs of Staff, in coordination with the combatant commands and Services. Also called **JP**. (JP 1-02)

leadership. The art and science of influencing and directing people to accomplish the assigned mission. (AFDD 1-1)

mission. 1. The task, together with the purpose, that clearly indicates the action to be taken and the reason therefore. 2. In common usage, especially when applied to lower military units, a duty assigned to an individual or unit; a task. 3. The dispatching of one or more aircraft to accomplish one particular task. (JP 1-02)

operational level of war. The level of war at which campaigns and major operations are planned, conducted, and sustained to accomplish strategic objectives within theaters or areas of operations. Activities at this level link tactics and strategy by establishing operational objectives needed to accomplish the strategic objectives, sequencing events to achieve the operational objectives, initiating actions, and applying resources to bring about and sustain these events. These activities imply a broader dimension of time or space than do tactics; they ensure the logistic and administrative support of tactical forces, and provide the means by which tactical successes are exploited to achieve strategic objectives. See also strategic level of war; tactical level of war. (JP 1-02)

specialty training. The total training process (life cycle) used to qualify Airmen in their assigned specialty. (AFI 36-2201) [AFDD 1-1]

strategic level of war. The level of war at which a nation, often as a member of a group of nations, determines national or multinational (alliance or coalition) security objectives and guidance, and develops and uses national resources to accomplish these objectives. Activities at this level establish national and multinational military objectives; sequence initiatives; define limits and assess risks for the use of military and other instruments of national power; develop global plans or theater war plans to achieve these objectives; and provide military forces and other capabilities in accordance with strategic plans. (JP 1-02)

tactical level of war. The level of war at which battles and engagements are planned and executed to accomplish military objectives assigned to tactical units or task forces. Activities at this level focus on the ordered arrangement and maneuver of combat elements in relation to each other and to the enemy to achieve combat objectives. (JP 1-02)

task force. 1. A temporary grouping of units, under one commander, formed for the purpose of carrying out a specific operation or mission. 2. A semi-permanent organization of units, under one commander, formed for the purpose of carrying out a continuing specific task. (JP 1-02)

training. Instruction and study focused on a structured skill set to acquire consistent performance. Training has predictable outcomes and when outcomes do not meet expectations, further training is required. (AFDD 1-1)

war. Open and often prolonged conflict between nations (or organized groups within nations) to achieve national objectives. (AFDD 1)

war game. A simulation, by whatever means, of a military operation involving two or more opposing forces using rules, data, and procedures designed to depict an actual or assumed real life situation. (JP 1-02)

IC 06-1 TO AFDD 1-1, *LEADERSHIP AND FORCE DEVELOPMENT*

18 FEBRUARY 2006

SUMMARY OF REVISIONS

These changes incorporate interim change (IC) 06-1. This change corrects/updates discussions on operational risk management as a responsibility of leadership (page 3); language addressing longer term force sustainment efforts within force development (page 24); and adds acronym ORM to Glossary (page 72).

www.ingramcontent.com/pod-product-compliance
Lightning Source LLC
Chambersburg PA
CBHW051351200326
41521CB00014B/2540